The Mind, The Brain, and Complex Adaptive Systems

The Mind, The Brain, and Complex Adaptive Systems

Editors

Harold Morowitz
George Mason University
Fairfax, VA

Jerome L. Singer
Yale University
New Haven, CT

Proceedings Volume XXII

Santa Fe Institute
Studies in the Sciences of Complexity

Routledge
Taylor & Francis Group
New York London

Director of Publications, Santa Fe Institute: *Ronda K. Butler-Villa*
Publications Assistant, Santa Fe Institute: *Della L. Ulibarri*

First published 1994 by Westview Press

Published 2018 by Routledge
711 Third Avenue, New York, NY 10017, USA
2 Park Square, Milton Park, Abingdon, Oxon OX14 4RN

Routledge is an imprint of the Taylor & Francis Group, an informa business

Copyright © 1994 Taylor & Francis

This volume was typeset using TEXtures on a Macintosh II computer.

ISBN 13: 978-0-2014-0986-4 (pbk)
ISBN 13: 978-0-201-40988-8 (hbk)

About the Santa Fe Institute

The *Santa Fe Institute* (SFI) is a multidisciplinary graduate research and teaching institution formed to nurture research on complex systems and their simpler elements. A private, independent institution, SFI was founded in 1984. Its primary concern is to focus the tools of traditional scientific disciplines and emerging new computer resources on the problems and opportunities that are involved in the multidisciplinary study of complex systems—those fundamental processes that shape almost every aspect of human life. Understanding complex systems is critical to realizing the full potential of science, and may be expected to yield enormous intellectual and practical benefits.

All titles from the *Santa Fe Institute Studies in the Sciences of Complexity* series will carry this imprint which is based on a Mimbres pottery design (circa A.D. 950–1150), drawn by Betsy Jones. The design was selected because the radiating feathers are evocative of the outreach of the Santa Fe Institute Program to many disciplines and institutions.

Santa Fe Institute Series List

Lecture Notes Volumes in the Santa Fe Institute Studies in the Sciences of Complexity

Proceedings Volumes in the Santa Fe Institute Studies in the Sciences of Complexity

Contents

Harold Morowitz
George Mason University, 207 East Building, Fairfax, VA 22030

Preface

The Krasnow Institute for Advanced Study at George Mason University was established by a bequest from the late Shelley Krasnow. Having decided that the Institute would at first focus in the area of the cognitive disciplines, the directors decided to institute the program with a conference on **The Mind, the Brain, and Complex Adaptive Systems**. Because the Krasnow Institute regards the Santa Fe Institute as a model for advanced studies and because of an overlap in approach, the enthusiastic support of the Santa Fe Institute was obtained as joint sponsor of this activity.

An *ad hoc* committee was organized, and each member submitted a list of names of their choice of speakers for such a conference. A surprising overlap of views produced a list of eighteen leaders in various fields of cognitive disciplines and approaches to complexity and cognition. From this group, twelve distinguished scholars were available and agreed to participate in the symposium.

The meeting was held at George Mason University on May 24–26, 1993, and resulted in considerable discussion of the area of intersection of neurobiology, cognitive psychology, and computational approaches to cognition. The papers and extended abstracts in this volume reflect the intellectual foundation upon which the Krasnow Institute will work in the ongoing pursuit of the understanding of the items addressed in this meeting.

Professor Jerome Singer of Yale University chaired two discussion sessions among the speakers. These were vigorous and useful in trying to communicate between the various approaches. Prof. Singer coedits this book and presents his overview.

A number of individuals at Santa Fe Institute, the Krasnow Institute, and George Mason Universtiy have been specially helpful in organizng this meeting and its publication, and at the risk of omission, I wish to thank Steven Diner, Lydia Walls, Iris Knell, Mark Friedlander, Jr., L. M. Simmons, Jr., and Ronda Butler-Villa.

Jerome L. Singer
Yale University, Department of Psychology, P. O. Box 208205, New Haven, CT 06520-8205

Mental Processes and Brain Architecture: Confronting the Complex Adaptive Systems of Human Thought (An Overview)

In the 1980s and 1990s we are witnesses to a new paradigmatic shift in science. Theorists in many fields are moving away from linear, reductionist, simple cause-effect models toward confronting the challenges of complex adaptive systems. Such systems are found in fields as diverse as astrophysics and quantum mechanics, cellular biology and species evolution, archeology and economics, cerebral neurobiochemistry and cognitive psychology. With the emergence in the early 1980s of the Santa Fe Institute under the leadership of physicists such as George Cowan and Murray Gell-Mann and, more recently, with the formation of the Krasnow Institute for Advanced Study, we see (in keeping with a basic complex adaptive systems principle) the emergence of new kinds of educational and research structures. These settings are designed to train young scientists as well as senior researchers to conduct investigations in new approaches that can address the complexities that have hitherto baffled us.

How can such processes operate? In Waldrop's intriguingly personalized book *Complexity*, we find an account of how two scientists working at the Santa Fe Institute attended a computer simulation on the flocking of birds, a demonstration at the nearby Los Alamos research center. It stimulated an entirely new line of research on economics.[20] John Holland (one of the contributors to this volume) and Brian Arthur, a former Santa Fe Fellow and, later, a Professor of Population Studies and Economics at Stanford University, had watched Craig Reynolds' simulation of bird behavior (using computer creatures called "boids") with three principles: (1) maintaining minimum distances from other objects, (2) matching velocities with other "boids," and (3) moving toward the apparent central mass of proximal "boids." With just those three rules the "boids" on the computer screen, initially randomly scattered, moved inexorably into flocks that stayed together as they confronted various obstacles, sometimes splitting up to flow around obstacles and then coming together again. When one of the "boids" accidentally hit a pole, it seemed to flutter around as if "stunned and lost—then, darted forward to rejoin the flock as it moved on" (Waldrop,[20] pg. 242).

The two scientists argued over and over whether such flocking was a truly emergent property, not inherently given by the program albeit, of course, growing ultimately out of the three simple defining rules of "boid" behavior. Out of their discussion Arthur was impelled to reexamine the way economic models are constructed and eventually to begin the stimulating efforts to construct new models of economic behavior that grew not only from the mathematical variables of microeconomics but from forming models of miniature economic societies using *agents* who could learn from their errors and interact to form structures and sequences that might emerge from chaotic or nonlinear initial "behaviors" to produce realistic economic processors.

In this volume we will present a series of papers, some outlines of research programs, others pointing to needed research and even a few that move in the direction of showing how even the great complexities of human thought, on the one hand, and the nature of brain structure, on the other, might be modeled along rule-governed principles that could evoke the emergent properties of complex adaptive systems. To provide a sense of order to the somewhat diverse nature of the contributions herein on mind and brain, we must first look again at the generally agreed upon principles of complex adaptive systems as outlined in the chapter by John Holland (one of the "boid" watchers in the example drawn from Waldrop's anecdote). The reader will also find a somewhat more complex elaboration of Holland's approach in Gell-Mann's sweeping chapter. To ease the reader into this volume and especially to the special properties of mind, however, I will draw on Holland's more concise approach.

Complex adaptive systems, first, all involve numerous interacting *agents* whose *aggregate behaviors* are to be understood. Such aggregate activity is *nonlinear*; hence, it cannot simply be derived from summation of individual components behaviors. These agents are *morphologically* diverse (as in the different forms of brain neurons and networks or their electronic and chemical transmission systems). Removal of one agent type leads the system to self-reorganization and a series of changes designed to make up for the gap in the system. There is a continuous evolving quality to the structure leading to the emergence of new agents or new interactions. Such novelty, as Holland notes, is a problem for mathematical analyses of the more traditional kind. Finally, the agents are characterized by *internal models*, built in rule-governed procedures that allow for the anticipation of consequences. An example from the psychology of learning is the way that reductionist stimulus-response models of rat learning gave way to conceptions of "plans" or cognitive maps that seemed to provide better explanations and predictions of learning.[18] Gell-Mann, in his chapter, pays special attention to "schemata" as properties of all complex systems. As we shall see in the research on human memory the construct of a schema as a central organizing structure, has opened the way for considerable advances in both experimental research and in computer simulation of mental processes. In the chapters in this volume by Goldman-Rakic and by Squire, such concepts of internal models derived from the more molar studies of intact organisms or from human thought can also be applied to understanding brain structure and localization of function.

COGNITION, CONSCIOUSNESS, AND COMPLEXITY

The contributors to this volume came together in the Krasnow Institute Conference without a clear intent of integrating their own scientific areas or lines of research into the definition of complex systems just presented. It is possible to suggest, however, that if we address certain features of human thought and consciousness from the standpoint of the seven defining steps outlined by Holland, we can identify areas of inquiry in modern psychology and cognitive science that reflect these principles. We can then take the next step of pointing to ways in which the chapters of this volume relate to the paradigm of complexity in science.

Consider the problems of human consciousness and its relation to the mind-body problem. A recent, very technical meeting at the Ciba Foundation in London brought together some of these same contributors (e.g., Harnad, Dennett, Kihlstrom, Singer) along with other philosophers, psychologists, and neuroscientists to search for possible integrative approaches to mental experience and underlying brain processes.[4] In the view of this participant, that meeting, however valuable the individual papers, suffered from a failure to search for a defining conception such as that afforded by complex adaptive systems.

Let us consider some issues of human thought and consciousness in relation to the definition derived from Holland. There seems increasingly to be agreement among researchers of human information acquisition and processing that we can describe our perceptual, cognitive, and memory functions through the use of sets of rule-following agents such as feature-detectors, working memory structures, schemata, scripts, and plans. These operate, however, within a broader system in which we receive signals and information not only from a consensually agreed upon *outside* world but also from sources of stimulation within our bodies. These include the twitches or tugs of our muscles, the rapidity of heart rate or blood pressure, gurglings or pains in our digestive tracts *and* also from the presumably ongoing workings of our brain. The brain machinery runs silently but it generates memories, daydreams, anticipatory images. In effect we are almost continuously in a situation in which we must assign priorities to the signal sources to which we will attend in given social situations since, however well we be able to process material in parallel, we must ultimately use sequential processing in critical perceptions and actions. We can drive long stretches on a relatively empty, well-known highway while listening to music and engaging in fantasies but, with a sudden build-up of traffic, we must revert to a relatively sequential mode lest we risk an accident. Simon's chapter in this volume makes a strong case for the necessity of a sequential view. Our various information-processing agents confronted with the varied sources of external, body-machinery and centrally generated signals almost certainly operate in a parallel nonlinear fashion in filtering cues and shifting attention. Sudden environmental changes, sudden bodily pains, sudden recurrent stressful memories, or the awareness of unfinished tasks may all compete simultaneously for attention, reflection, and action.

My emphasis on this complexity points the way also to still another morphologically distinct system, our emotions. These can be aroused either through cognition or, as is increasingly clear, they may also follow the separate pathways of pure conditioned responses.[10] There are good theoretical and empirical reasons for identifying emotions as morphologically distinct agents from the cognitive processors emphasized in the present volume.[9,17,19] Despite their distinctive and differentiated physiological and psychological properties, however, the emotions are subtly interwoven with our information processing. While they can be evoked by pure associative conditioning, they are also largely responsive to interruptions and mismatches in our ongoing efforts to organize and to integrate information from the three types of signal sources described above into meaningful schemata drawn from working and long-term memory.[11,14,19]

I have mentioned the emotions here in order to stress the complex mix of agents to be considered in recognizing the challenge posed by psychological systems with conscious or unconscious representation for defining models that can be mapped onto (a top-down approach) or defined (a bottom-up approach) by our exciting new understandings of brain structure and function (see chapters by Churchland, by Goldman-Rakic, and by Squire and Knowlton). This volume focuses primarily on our cognitive functions. A full grasp of human consciousness or of its presumed neural underpinning must ultimately address how the complex interactions between acquired schemata or scripts, anticipations and the mismatches or ambiguities in processing novel information immediately or over prolonged time periods, evoke specific affective states. The very different time cycles of emotional responses once aroused and of cognitive processes—the former slow and prolonged, creating a network of modifying body changes, the latter operating in milliseconds of associations—provide a difficult problem for standard mathematical formulation.

Holland's definition of complex adaptive systems next refers to reorganizations and gap filling. We know that despite the nonregeneration of damaged neural tissue and the consequent permanent loss of certain psychological capacities associated with brain damage (see Squire and Knowlton's chapter), humans and some animals can make remarkable adaptations to sustain their cognitive and social adaptation. Some of the most elegant and impressive neurological and psychological work produced in the first half of this century was that of Kurt Goldstein. He demonstrated how seriously impaired brain-injured adults found new ways of reading, of guiding themselves around their environment, or of communicating to actualize their potentialities.[6,7] While the psychological reality of such gap-filling is well established, the neurological processes that presumably are correlated with or underlie such adaptations are far more difficult to ascertain. Perhaps the new imaging technology pointed to in the chapters by Churchland and by Goldman-Rakic can contribute to our understanding in the cases of the grossly brain injured.

But there are far more subtle forms of demonstrable conscious or unconscious human reorganizational activities that pose a challenge to current brain architectural and functional knowledge. Our vast human capacity for outright lying, for self-deception and rationalization, for misremembering or intentionally forgetting (especially in social situations that are self-defined as threatening) seems as yet far beyond explanation through the bottom-up neural structures described by Churchland in her chapter's support of Crick's hypotheses.[15] Indeed such processes (which we can now witness occurring day after day on Court TV, if not in our own universities, businesses, or governments) are scarcely addressed by the excellent chapters herein which focus primarily upon direct experiences of the external world. There are hints, however, in the chapters by Kihlstrom, by Dennett, by Schank, and by Antrobus of how we might formulate such processes and even model these self-generated virtual realities. Our human ways of reconstructing "physically given" realities continue to reflect an especially difficult challenge posed by the mind as a complex adaptive system.

Implicit in the issue of our capacities for recording and modifying our representations of physical and social realities are two other principles of Holland, the emergent characteristics and the central rule-providing models of complex systems. Considerable progress has been made in psychology and cognitive science in the past thirty years through the identification and definition of structures of memory such as *episodic* and *event memory*, *working memory*, *implicit* and *explicit memory*, *short-*, and *long-term memory* and those organized mental representations that serve to strengthen (and often to bias) encoding and retention of new material: *schemata*, *scripts*, *prototypes*, or *narratives*. The chapters by Goldman-Rakic and by Squire point to links between the memory structures and brain areas and pathways while the chapters by Kihlstrom, by Rumelhart, by Antrobus, and by Schank contribute to our understanding of organized mental structures.

Modeling the emergent characteristics of the mind presents probably the most difficult task for creating links between the hardware or software of human biology and the achievements of human consciousness. Our most dramatic examples of creativity, the formulation of grand and testable theories by Newton, Darwin, and Einstein, the deaf Beethoven composing the rich harmonies of the *Quartet in E-Flat* or of the *Grosse Fuge* are only extremes of the emergent productions that humans manifest daily. From the preschool child playing at make-believe, modeling a town or a strange country with just a few blocks, to the daydreams of adolescents and the nocturnal dreaming that we all engage in, creativity is a central feature of the human condition.[12] What of the human tendency to create self-representations that can then become guides for our emotional experiences, for our patterns of memory, and for experiences as varied as forms of pathology, immune system response, or the capacity for self-reconstruction in the face of adversity?[3,8,13,16] Some suggestions on modeling of such self-representations can be found in the chapters by Kihlstrom, by Antrobus, and by Schank. Once again the links to the brain are not obvious. There are suggestions in the references to the breakdown of certain representational and sequential capacities in the chapters by Churchland and by Goldman-Rakic, and in Squire and Knowlton's references to Alzheimer's that offer some clues here.

We have so far focused on consciousness and mind as complex systems without emphasizing the social interpretive or social interaction dimensions of human experience. Consider the inherent complexity introduced by Apter's *reversal theory* which is supported by a good body of empirical research.[1,2] He proposes that humans regularly shift between two states of mind in which arousal can be experienced in opposite ways. When we are goal-focused and "working" towards a clear objective, high arousal can be anxiety provoking and disruptive. When we are in a self-defined play or nontelic set, high arousal can be pleasant and low arousal can be boring. We move through each day fluctuating between states and reversing our interpretations of social events or of our arousal experiences. If we "frame" events as playful (as in nonprofessional sports or as in movie-going), the high arousal produced by a real of vicarious risk (a horror film) can be taken as pleasant. Quite different experiences and actions follow depending on whether we use the play or nontelic frame.

I mention this viewpoint which has broad ramifications in social and personality psychology because it once again points up the creation of virtual realities that may defy easy "bottom-up" physiological explanation on the basis of our current biological knowledge. Churchland in her chapter questions Dennett's necessity for construction of a "software virtual reality" between experienced consciousness and brain structure and Dennett, albeit focusing in his chapter on evolution, seeks to sustain his views as expressed in *Consciousness Explained.*[5] The reader of this volume can decide whether Churchland's emphasis on parallel distributed processing and her example drawn from Crick's work provides a solid basis for a "bottom-up" approach to explaining the kinds of mental phenomena cited above.

The excitement reflected since the mid-1980s by the concepts of neural nets and of parallel distributed processing is confronted again in this book. Simon, as a pioneer in computer simulation and artificial intelligence, challenges the broad value of such models on several grounds in his chapter. In their respective contributions Rumelhart and Churchland support the utility of the PDP metaphor as bringing us closer to a computer analogue that seems at least plausibly relevant to the actual operation of networks of interconnected neural fibers and pathways in the "wet" brain. Churchland goes so far as to point, albeit briefly, to a phenomenon seemingly so remote as daydreaming. This theme is picked up in much greater detail by Antrobus, in his chapter, which begins with a systematic examination of how our stream of consciousness shifts between task-related processing and "task-unrelated intrusive thought" (TUITs), our fleeting daydreams, fantasies, andshifts of attention toward anticipated futures. He then moves to examining how such empirically studied phenomena can be modeled in terms of PDP computer simulations. Then on the basis of research findings suggesting the continuities between waking daydreams and dream reports obtained in sleep laboratories, he presents an effort to model some of the properties of night dreams using the methods and assumptions of parallel processing in the brain.

To provide a first guidepost for the reader, then, this volume opens with a broad overview across the sciences of complex adaptive systems offered by Gell-Mann. Simon then moves to an examination of one phase of the such systems, their complex hierarchical structures. He also deals especially with human and machine-simulated expertise focusing primarily on sequential and logical thought. Holland and Dennett examine ways of modeling the evolutionary process in general terms. Harnad raises questions about computer modeling of complex human thought, especially the more subtle features of experience and points to some boundary conditions or constraints that must be considered in trying to reconcile empirical psychological research, computer simulation, and the presumed actual workings of the nervous system and brain. Churchland, taking a firm stand against any hints of dualism, provides an optimistic assessment of the value of a PDP and bottom-up approach to understanding mental phenomena via the combination direct brain experimentation or and neural net modeling. Goldman-Rakic outlines a step-by-step elegant research program for, in effect, localizing the architecture and pathways of working memory in the monkey brain. She carefully points to problems that need to be explored from a complex adaptive systems perspective. Squire and Knowlton review the human brain damage literature to support those important *agents* of memory, implicit and explicit, as well as many other properties of our retention and reuse of learned material. Kihlstrom uses these agents of memory, and draws on basic cognitive research, social and personality psychology experiments, and the phenomenon of hypnosis. He tries to demonstrate that a model of unconscious cognitive processing is necessary to understand conscious thought and behavior but in a very different way than the diffuse and vague mechanisms described by Freud and Jung.

Rumelhart and Antrobus both provide vivid exemplifications of parallel distributed modeling approaches to complex features of human thought. Finally, Schank using his own conception of the important role of the narrative process or "storytelling" as a fundamental agent in human thinking and learning moves us out of the laboratory and points towards the practical educational potential of his approach. Computer simulation of storytelling and direct teaching can go hand-in-hand to improve learning.

Often enough a conference of this type can well exemplify nonlinearity and chaos as properties of human social gatherings. Without denying the range and diversity of approaches presented herein, I believe there are significant threads bearing on new ways of looking at data and modeling human thought that will provoke and excite the reader. Rarely have so original and impressive a group of scholars and empirical researchers come together for so thoughtful a symposium.

REFERENCES

1. Apter, M. *Reversal Theory: Motivation, Emotion and Personality*. London & New York: Routledge, 1989.
2. Apter, M. *The Dangerous Edge*. New York: The Free Press, 1993.
3. Avants, S. K., J. L. Singer, and A. Margolin. "Self-Representations and Negative Affect in Cocaine-Dependent Individuals." *Imagination, Cog. & Pers.* **13(1)** (1993-94): 3–24.
4. Ciba Foundation. *Experimental and Theoretical Studies of Consciousness*. Symposium # 174. New York: Wiley, 1993.
5. Dennett, D. *Consciousness Explained*. Boston, MA: Little, Brown, 1991.
6. Goldstein, K. *The Organism*. New York: American Book Co., 1939.
7. Goldstein, K. *Human Nature in the Light of Psychopathology*. Cambridge, MA: Harvard University Press, 1940.
8. Higgins, E. T. "Self-Discrepancy: A Theory Relating Self and Affect." *Psychol. Rev.* **94** (1987): 319–340.
9. Izard, C. *The Psychology of Emotions*. New York: Plenum, 1991.
10. LeDoux, J. E. "The Neurobiology of Emotion." In *Mind and Brain: Dialogues in Cognitive Neuroscience*, edited by J. E. LeDoux and W. Hirst, 301–354. New York: Cambridge University Press, 1988.
11. Mandler, G. *Mind and Body*. New York: Norton, 1984.
12. Singer, D. G., and J. L. Singer. *The House of Make-Believe: Children's Play and the Developing Imagination*. Cambridge, MA: Harvard University Press, 1990.
13. Singer, J. A., and P. Salovey. *The Remembered Self: Emotion, Memory and Personality*. New York: The Free Press, 1993.
14. Singer, J. L. *The Human Personality*. New York and San Diego: Harcourt Brace Jovanovich, 1994.
15. Singer, J. L. *Repression and Dissociation: Implications for Personality, Psychopathology and Health*. Chicago, IL: University of Chicago Press, 1990.
16. Strauman, T. J., A. Lemineux, and C. Coe. "Self-Discrepancy and Natural Killer Cell Activity: Immunological Consequences of Negative Self-Evaluation." *J. Pers. & Soc. Psychol.* **64** (1993): 1042–1052.
17. Thompson, J. *The Psychobiology of Emotions*. New York: Plenum, 1988.

18. Tolman, E. C. "The Determiners of Behavior at a Choice Point." *Psych. Rev.* **45** (1938): 1–41.
19. Tomkins, S. S. *Affect, Imagery, Consciousness.* Vol. 1. New York: Springer-Verlag, 1962.
20. Waldrop, M. *Complexity. The Emerging Science at the Edge of Order and Chaos.* New York: Simon & Schuster, 1992.

Murray Gell-Mann
Santa Fe Institute, 1399 Hyde Park Road, Santa Fe, NM 87501 and Los Alamos National Laboratory, Los Alamos, NM 87545

Complex Adaptive Systems

This chapter originally appeared in *Complexity: Metaphors, Models, and Reality*, edited by George Cowan, David Pines, and David Meltzer (Reading, MA: Addison-Wesley, 1994). Copyright © Addison-Wesley; reprinted by permission.

INTRODUCTION

The various groups in the Santa Fe Institute family studying complex adaptive systems (CAS) have somewhat different points of view and have adopted different vocabularies. Some of us speak of "artificial life" or "artificial social life" or "artificial worlds," while others, of whom I am one, prefer to consider natural CAS and computer-based systems together. The latter include methods for adaptive computation as well as models and simulations of natural CAS.

Even the term CAS has different meanings for different researchers. As one distinguished professor at this conference remarked, "a scientist would rather use someone else's toothbrush than another scientist's terminology." For example, my nomenclature differs from that of John Holland, from whom I have learned so much.

He calls something a CAS only if it is a collectivity of interacting adaptive agents, each of which I would refer to as a CAS. Likewise, John uses the term "internal model" to mean what I call a schema.

There are additional possible sources of misunderstanding as well, stemming from the relation between computer-based and natural systems. At one of our Science Board Symposia, a speaker asked, "Are we using computation as an aid in understanding biology (e.g., evolution, thinking, etc.) or are we using biology as a metaphor for work on computation?" That is an important question. At some institutions where computation and neural systems are studied, there is real confusion on this issue. For example, success in designing a computing system based on "neural nets" is sometimes taken as evidence that such nets furnish a serious model of the human brain, with the units or nodes corresponding to individual neurons.

I favor a comprehensive point of view according to which the operation of CAS encompasses such diverse processes as the prebiotic chemical reactions that produced life on Earth, biological evolution itself, the functioning of individual organisms and ecological communities, the operation of biological subsystems such as mammalian immune systems or human brains, aspects of human cultural evolution, and adaptive functioning of computer hardware and software. Such a point of view leads to attempts to understand the general principles that underlie all such systems as well as the crucial differences among them. The principles would be expected to apply to the CAS that must exist on other planets scattered through the universe. Most of those systems will of course remain inaccessible to us, but we may receive signals some day from a few of them.

As to successful adaptive computational methods and devices, we have examples such as neural net systems, based on a perceived similarity, even though it may be rather remote, to the functioning of the human brain, and genetic algorithms, based on a resemblance to evolutionary processes. Surely these sets of methods belong, together with many others, mostly as yet undiscovered, to a huge class of computational CAS, with common features that will be well worth identifying and understanding. Some of the new computational methods may exhibit similarities to the operation of natural CAS that we know, such as the immune system, but others may be quite unlike any natural process familiar to us.

A CAS gathers information about its surroundings and about itself and its own behavior, at a certain level of coarse graining. The time series that represents this information can sometimes be approximated by a steady one, although in general it is changing with time, frequently in ways that depend on the system's behavior, and the surroundings are often coevolving. The following[1,2] are general characteristics of a CAS:

1. Its experience can be thought of as a set of data, usually input → output data, with the inputs often including system behavior and the outputs often including effects on the system.
2. The system identifies perceived regularities of certain kinds in the experience, even though sometimes regularities of those kinds are overlooked or random

features misidentified as regularities. The remaining information is treated as random, and much of it often is.

3. Experience is not merely recorded in a lookup table; instead, the perceived regularities are compressed into a schema. Mutation processes of various sorts give rise to rival schemata. Each schema provides, in its own way, some combination of description, prediction, and (where behavior is concerned) prescriptions for action. Those may be provided even in cases that have not been encountered before, and then not only by interpolation and extrapolation, but often by much more sophisticated extensions of experience.

4. The results obtained by a schema in the real world then feed back to affect its standing with respect to the other schemata with which it is in competition.

Now the feedback process need not be a clear-cut one in which success is well defined and leads to survival of the schema while failure, equally well defined, results in its disappearance. Fitness may be an emergent or even an ill-defined feature of the process; the effect on the competition among schemata may be only a tendency; and a demoted schema may be kept for use in a subordinate capacity or retained in memory while not utilized (it might, after all, produce useful variants). The important thing is the nature of the selection pressures exerted in the feedback loop, whether or not they are expressible in terms of a fitness function. (Similarly, physical forces may or may not be derivable from a well-defined potential.)

An excellent example of a CAS is the human scientific enterprise, in which the schemata are theories, giving predictions for cases that have not been observed before. There is a tendency for theories that give successful predictions (and exhibit coherence with the body of successful theory) to assume a dominant position, although that is by no means a simple, mechanical procedure. Older, less successful theories may be retained as approximations for use in restricted sets of circumstances. Even wrong theories are not necessarily wholly forgotten, since they may inspire some useful theoretical work in the future.

In its application to the real world, a schema is in a sense reexpanded, reequipped with some of the arbitrariness of experience, some of the random material of the kind that was stripped away from the data when regularities were identified and compressed. For instance, a theory must be combined with boundary conditions in order to give a prediction. The additional data adjoined to the schema may simply be part of the continuing stream of incoming data, which contain, in general, the random along with the regular.

In most CAS the level of the schemata and the level at which results are obtained in the real world are entirely distinct. In the realm of biological organisms, that is the distinction between genotype and phenotype, where the phenotype depends not only on the genotype but on all the accidents of development that intervene between the DNA and the adult organism. However, in some cases, such as Tom Ray's world of digital organisms, the genotype and phenotype are not physically different, but distinguished only by function. His sequences of machine instructions play both roles. As Tom Ray remarks, certain theories of the origin of life on Earth assert that RNA once behaved that way, both as bearer of information and as agent of chemical activity, before the appearance of organisms exhibiting separate genotype and phenotype.

Some new computer simulations of evolution try to include distinct genotypic and phenotypic levels. One that is under development at UCLA even simulates sexual reproduction, with haploid and diploid generations, and tries to test William Hamilton's idea that the principal utility of the male lies in helping to outrace enemies, especially parasites, by providing the offspring with genetic diversity that would be lacking in parthenogenesis.

Complex adaptation is to be contrasted with simple or direct adaptation, as in a thermostat, which just keeps mumbling to itself, "It's too cold, it's too cold, it's too hot, it's just right, it's too cold," and so forth. In the 1940s, the chemist Cyril (later Sir Cyril) Hinshelwood put forward a direct adaptation theory of the development of bacterial resistance to drugs. Genetic variation and selection were rejected in favor of a straight negative feedback process in chemical reactions in the cell. The drug interfered at first with the chemistry of the cell, but then the deleterious effects were mitigated as a result of reaction dynamics, and the mitigation was transmitted mechanically by the bacteria to their progeny in the course of cell division. There was no compression of regularities, no competition of schemata.

Hinshelwood's theory lost out, of course, but it has not been totally forgotten, and it now serves my purpose as an example of direct adaptation rather than the operation of a CAS. Direct control mechanisms are common in nature and in human industry, and they formed the subject matter of cybernetics half a century ago.

The cybernetic era was followed by the era of the expert system, employing a fixed "internal model" designed using the advice of experts in a field, for instance a decision tree for medical diagnosis. The expert system did not learn from the results of its work, however. It remained fixed until it was redesigned. (Only if the human redesigners are included can the expert system be regarded as a CAS, of the kind that involves "directed evolution" or "artificial selection," with humans in the loop.) The new era of CAS in robotics and other such fields is the age of constructed systems that actually learn, by formulating schemata subject to variation and to selection according to results in the real world.

It is useful to distinguish various levels of adaptation. In particular, we can take the example of human societies, where a schema is a set of customs, traditions, myths, laws, institutions, and so forth, what Hazel Henderson calls "cultural DNA." (The biologist Richard Dawkins has invented the word "meme" for a unit of that DNA analogous to a gene.)

The schemata include prescriptions for collective behavior. A culture operating on the basis of a given schema reacts to altered circumstances such as climatic change, invasion, and so forth, in ways prescribed by that schema. If the climate turns warmer and drier, the response of a group of villages may be to move to higher elevations. In the event of attack by outsiders, the inhabitants of all the villages may retire to a fortified site, stocked with food and water, and sustain a siege. What happens at this level is something like direct adaptation.

On the next level, the society may change its schema when the prevailing one does not seem to have given satisfactory results. Instead of migration to the high-lands, the villagers may try new crops or new methods of irrigation or both. Instead of retreating to a fort, they may respond to invasion with a counterattack aimed at the enemy's heartland.

Finally, there is the level of Darwinian survival of the fittest (as in population biology). In some cases, not only does a schema fail, but the whole society is wiped out. (The individual members need not all die, but the society ceases to exist as a functioning unit.) At this level the successful schemata are the ones that permit the societies using them to survive.

Not only are these three levels of adaptation distinct, but the time scales associated with them may be very different. Nevertheless, discussions of adaptation in the social science literature sometimes fail to discriminate among the levels, with unfortunate results for clarity.

The disappearance of societies is somewhat analogous to the death of organisms or to the forgetting of ideas. Such phenomena are, of course, universal and not unrelated to the second law of thermodynamics. Still, over a given period of time, the importance of mortality can vary from one domain to another.

In cases where death is very important at the phenotypic level, a crucial measure of success for a schema is phenotypic survival, and reproduction assumes great significance. Moreover, population can then supply a rough quantitative measure of fitness. In biology, one often follows the population of a cluster of genotypes such as a species or subspecies, and the clustering phenomenon is itself of very great interest. One can also follow subpopulations characterized by particular alleles of certain genes.

By contrast, there are situations where death is comparatively unimportant, whether at the genotypic or the phenotypic level. One schema can dominate an-other without the losing one disappearing; reproduction is not of overwhelming interest; and population is not of critical importance as a measure of fitness. Consider individual human thinking, for example. If we try to grasp an issue more clearly than before, we may succeed in getting an idea that dispels a great deal of previous confusion and displaces, to a considerable extent, earlier ideas. (That is not so easy, by the way, because existing ideas entrench themselves and we have a tendency to interpret new information as confirmatory, so that we dig ourselves deeper and deeper into what may be a quite unsuitable hole.) Over time scales such that forgetting is not a crucial factor, replication and population are not particu-larly relevant concepts to the success of an idea in the thinking of an individual person. What matters most is that at the real-world level one idea has received more positive feedback than another and thus assumed a comparatively dominant position. Over a very long time scale, of course, every system eventually has to get rid of clutter in some way, so that erasure, forgetting, or some other kind of grim reaper has to come into the picture.

Looking at CAS overall, we see that fitness is a rather elusive concept when it is endogenous. If an exogenous criterion is supplied, as in a machine that is designed and programmed to win at chess, then of course the feedback loop involves a well-defined fitness. But when fitness is emergent, it is not so easy to define without a somewhat circular argument in which whatever wins is fit by definition, and whatever is fit is likely to win.

As everyone recognizes, fitness is even less well defined when it is acknowledged that the surroundings of the system are themselves undergoing change and often coevolving. In the latter case, fitness "landscapes," even to the extent that they could be defined for fixed surroundings, now give way to a picture of shifting and interdependent landscapes for the different adaptive components of the total system.

The greatest difficulty in discussing features of a system that are "adaptive" (or that render it "fit") is the distinction between what is adaptive and what has resulted from a process of adaptation. The latter may often be maladaptive. Let us discuss some common reasons for that.

The simplest reason is, of course, that a CAS engages, under the influence of selection pressures in the real world, in a search process over the abstract space of schemata that is necessarily imperfect. Even if fitness is well defined, a system that merely searches for local maxima by "hill climbing on a landscape" would most often get stuck on a molehill. To have a chance to find mountains nearby, the search process must include other features, such as noise (but not too much noise) or else pauses in climbing to allow for free exploration. Naturally, schemata that are more or less maladaptive are often selected.

Apparently maladaptive schemata often occur for another reason, namely that the system is not defined broadly enough to encompass all the important selection pressures that are operating on the schemata concerned. For example, in the scientific enterprise, it would be a mistake to ignore the pressures other than purely scientific ones that affect the viability of a schema, especially in the short run. Scientists often exhibit human frailty, and issues of jealousy, greed, and the misuse of power may play a role in the fate of theories; even observational data are occasionally falsified. Of course it is equally foolish to exaggerate the importance of these extra-scientific selection pressures and to ignore the powerful correcting effect that comparison with nature keeps supplying.

The prevalence of prescientific theories, such as those associated with sympathetic magic, provide even more striking examples of the breadth of selection pressures. Suppose the members of a tribe believe in the efficacy of bringing rain by pouring out on the ground water obtained in a special place in the mountains. Clearly it is not carefully controlled comparison with results that sustains faith in the procedure, but selection pressures of very different kinds. For instance, the authority of powerful individuals or groups may be enhanced by the prevalence of belief in the ceremony, which may, in addition, be part of a whole set of customs that cement the bonds holding the society together.

More generally, it is significant that any CAS is a pattern-recognition device that seeks to find regularities in experience and compress them into schemata. Often it will find fake regularities where there is in fact only randomness. A great deal of superstitious belief can probably be attributed simply to that effect, which might be labeled the "selfish schema." (I have already mentioned how new data are often interpreted so as to strengthen an existing belief.)

Of course, a CAS will often err in the other sense and overlook regularities. Both types of error are presumably universal. In the realm of human beliefs, overlooking obvious regularities can usually be identified with denial. It is striking that in human beings both superstition and denial are typically associated with the alleviation of fear: in the former case fear of the random and uncontrollable and in the latter case fear of regularities that are all too evident, like the certainty of death.

Another example of the breadth of selection pressures comes up in studying the evolution of human languages. Here one should first of all distinguish several different CAS, at different levels and on different time scales. One is the evolution, over hundreds of thousands or millions of years, of the biological capacity to use languages of the modern type. Another is the evolution of those languages themselves, over thousands or tens of thousands of years. Yet another is the learning of a native language by a child. Consider the second of these three systems, concentrating for example on the evolution of grammar and phonology. One encounters, of course, the usual mixture of fundamental rules (in this case the "innate" constraints on grammar and phonology determined by biological evolution), frozen accidents or founder effects (in this case arbitrary choices made in ancestral languages that may have been transmitted to their descendants), and what is adaptive (in this case features that make for more effective communication). However, the selection pressures in linguistic evolution are not wholly linguistic. A great deal depends on whether a people speaking one language is more advanced culturally or stronger militarily than a people speaking another language. Such matters may easily have a greater effect on the fates of the two languages than which one is more convenient for communication.

Another common reason why maladaptive features arise from a process of adaptation is that time scales are mismatched. When circumstances change much more rapidly than the response time of the CAS, traits occur that may have been adaptive in the past but are so no longer. For instance, global climate change on a scale of a few decades will not permit the same kind of ecological adaptation that would be possible in the case of much slower change.

The human tendency to form groups that don't get along with one another, based on what are sometimes rather minute differences that an outsider would barely perceive, may be to a considerable extent an inherited tendency, even though it is fortunately subject to modification through culture. If a hereditary component is really involved, it may have been adaptive under the conditions that prevailed many tens of thousands of years ago. For example, it could have served to limit the size of the population in a given area to a number that the area could support. Nowadays, in a world of destructive weapons, the tendency seems quite maladaptive.

The phenomenon of imprinting provides an extreme case of the mismatch of time scales. A greylag goose that glimpsed Konrad Lorenz instead of its mother when it was first hatched was condemned to treat Lorenz as its mother ever after. The process of imprinting, which works fine in the more common case when the gosling sees its real mother, compromised forever the chances of a normal goose life for any gosling that saw Lorenz instead.

A milder phenomenon is that of windows of maturation. Béla Julesz emphasizes that certain abnormalities in vision have to be corrected early in childhood if they are to be corrected at all. In the case of learning deficits, it is important for public policy to know the extent to which they must be remedied during the first two years or so of life and the extent to which plasticity of the central nervous system permits them to be dealt with later by such programs as Head Start. (Of course the chances of success of Head Start are in any case compromised if the duration and intensity of the program are insufficient, as they often are.)

We must pay attention to time scales for other reasons as well. Fundamental rules on one scale of space and time may reveal themselves to be the results of frozen accidents on a larger scale. Thus the rules of terrestrial biology (such as the occurrence of DNA based on the nucleotides abbreviated A, C, G, and T) may turn out to represent just one possibility out of very many. On a cosmic scale of space and time, the earthly rules would then have the character of a frozen accident or founder effect. That is already widely believed to be the case for the occurrence of certain right-handed molecules in important biological contexts where the corresponding left-handed molecules do not occur. (Attempts to derive that asymmetry from the left-handedness of the weak interaction for matter, as opposed to antimatter, do not seem to have succeeded.)

Some of the most interesting questions about CAS have to do with their relations to one another. We know that such systems have a tendency to spawn others. Thus biological evolution gave rise to thinking, including human thought, and to mammalian immune systems; human thought gave rise to computer-based CAS; and so on. In addition, CAS are often subsystems of others, as an immune system forms part of an organism. Often, a CAS is a collectivity of adaptive agents, each a CAS in its own right, constructing schemata describing one another's behavior. One of the most important branches of the emerging science of CAS concerns the inclusion of one such system in another and the functioning of collectivities such as ecological communities or markets.

One class of composite CAS of particular interest consists of natural or computer-based systems with human beings "in the loop," as in the breeding of animals or plants (what Darwin called artificial selection as opposed to natural selection) or as in a computer system that creates pictures by presenting a human being with successive choices of alterations in an initial pattern.

Pure computer-based CAS can be used for adaptive computation, for modeling or simulating in a crude fashion some natural CAS, and for study as examples of CAS. In all three capacities, they illustrate that astonishingly great apparent complexity can emerge from simple rules, alone or accompanied by a stochastic process. It is always a fascinating and useful exercise to try to prune the rules, making them even simpler, while retaining the apparent complexity of the consequences. Such investigations will gradually lead to a mathematical science of rules and consequences, with theorems initially conjectured on the basis of examples and then proved.

Applications to natural or behavioral sciences require, at a minimum, not just those abstract propositions about rules and consequences but also additional information specifying situations simulating in some convincing way ones that arise in the science in question.

Still more information must be supplied if the computer model is to have any relevance to policy. Conditions prevailing on the planet Earth, including human institutions as well as features of the biosphere, have to be at least vaguely recognizable in the model. Even then, it is critical to use the results mainly as "prostheses for the imagination" in forecasting or in discussing policy options. Trying to fit policy matters into the Procrustean bed of some mathematical discipline can have most unfortunate consequences.

It is a major challenge to try to construct bridges connecting these different levels of abstraction, while maintaining the distinctions among them.

When we ask general questions about the properties of CAS, as opposed to questions about specific subject matter such as computer science, immunology, economics, or policy matters, a useful way to proceed, in my opinion, is to refer to the parts of the CAS cycle,

I. coarse graining,

II. identification of perceived regularities,

III. compression into a schema,

IV. variation of schemata,

V. application of schemata to the real world,

VI. consequences in the real world exerting selection pressures that affect the competition among schemata,

as well as to four other sets of issues:

VII. comparisons of time and space scales,

VIII. inclusion of CAS in other CAS,

IX. the special case of humans in the loop (directed evolution, artificial selection), and

X. the special case of composite CAS consisting of many CAS (adaptive agents) constructing schemata describing one another's behavior.

Here, in outline form, is an illustrative list, arranged according to the categories named, of a few features of CAS, most of them already being studied by members of the Santa Fe Institute family, that seem to need further investigation:

I. Coarse Graining

1. Tradeoffs between coarseness for manageability of information and fineness for adequate picture of the environment.

II. Sorting Out of Regularities from Randomness

1. Comparison with distinctions in computer science between intrinsic program and input data.
2. Possibility of regarding the elimination of the random component as a kind of further coarse graining.
3. Origin of the regularities in the fundamental laws of nature and in shared causation by past accidents; branching historical trees and mutual information; branching historical trees and thermodynamic depth.
4. Even in an infinite data stream, it is impossible to recognize all regularities.
5. For an indefinitely long data stream, algorithms for distinguishing regularities belonging to a class.
6. Tendency of a CAS to err in both directions, mistaking regularity for randomness and vice versa.

III. Compression of Perceived Regularities into a Schema

1. If a CAS is studying another system, a set of rules describing that system is a schema; length of such a schema as effective complexity of the observed system.
2. Importance of potential complexity, the effective complexity that may be achieved by evolution of the observed system over a given period of time, weighted according to the probabilities of the different future histories; time best measured in units reflecting intervals between changes in the observed system (inverse of mutation rate).
3. Tradeoffs between maximum feasible compression and lesser degree that can permit savings in computing time and in time and difficulty of execution; connection with tradeoffs in communication theory—detailed information in data base versus detailed information in each message and language efficiency versus redundancy for error correction.

4. Oversimplification of schema sometimes adaptive for CAS at phenotypic (real world) level.
5. Hierarchy and chunking in the recognition of regularities.

IV. Variation of Schemata

1. In biological evolution, as in many other cases, variation always proceeds step by step from what already is available, even when major changes in organization occur; vestigial features and utilization of existing structures for new functions are characteristic; are there CAS in which schemata can change by huge jumps all at once?
2. Variable sensitivity of phenotypic manifestation to different changes in a schema; possibility in biological case of long sequences of schematic changes with little phenotypic change, followed by major phenotypic "punctuations;" generality of this phenomenon of "drift."
3. Clustering of schemata, as in subspecies and species in biology or quasispecies in theories of the origin of life or word order patterns in linguistics—generality of clustering.
4. Possibility, in certain kinds of CAS, of largely sequential rather than simultaneous variants.

V. Use of the Schema (Reexpansion and Application to Real World)

1. Methods of incorporation of (largely random) new data.
2. Description, prediction, prescribed behavior—relations among these functions.
3. Sensitivity of these operations to variations in new data.

VI. Selection Pressures in the Real World Feeding Back to Affect Competition of Schemata

1. Concept of CAS still valid for systems in which "death" can be approximately neglected and reproduction and population may be correspondingly unimportant.
2. Exogenous fitness well-defined, as in a machine to play checkers; when endogenous, a elusive concept: attempts to define it in various fields, along with seeking maxima on "landscapes."
3. Noise, pauses for exploration, or other mechanisms required for the system to avoid getting stuck at minor relative maxima; survey of mechanisms employed by different systems.
4. Procedures to use when selection pressures are not derivable from a fitness function, as in neural nets with (realistic) unsymmetrical coefficients.
5. Possible approaches to the case of coevolution, in which the fitness concept becomes even more difficult to use.

6. Situations in which maladaptive schemata occur because of mismatch of time scales.
7. Situations in which maladaptive schemata occur because the system is defined too narrowly.
8. Situations in which maladaptive schemata occur by chance in a CAS operating straightforwardly.

VII,VIII. Time Scales; CAS Included in Others or Spawned by Others

1. Problems involved in describing interactions among CAS related by inclusion or generation and operating simultaneously on different levels and on different time scales.

IX. CAS with Humans in the Loop

1. Information about the properties of sets of explicit and implicit human preferences revealed by such systems.

X. CAS Composed of Many Coadapting CAS

1. Importance of region between order and disorder for depth, effective complexity, etc.
2. Possible phase transition in that region.
3. Possibility of very great effective complexity in the transition region.
4. Possibility of efficient adaptation in the transition region.
5. Possibility of relation to self-organized criticality.
6. Possible derivations of scaling (power law) behavior in the transition region.
7. With all scales of time present, possibility of universal computation for the system in the transition region.

ACKNOWLEDGMENTS

It is a pleasure to acknowledge the great value of conversations with John Holland and with other members of the SFI family. My research has been supported by the U.S. Department of Energy under Contract No. DEAC-03-81ER40050, by the Alfred P. Sloan Foundation, and by the U.S. Air Force Office of Scientific Research under the University Resident Research Program for research performed at Phillips Laboratory (PL/OLAL).

REFERENCES

1. Gell-Mann, Murray. "Complexity and Complex Adaptive Systems." In *The Evolution of Human Languages*, edited by M. Gell-Mann and J. A. Hawkins. Santa Fe Institute Studies in the Sciences of Complexity, Proc. Vol. X. Reading, MA: Addison-Wesley, 1992
2. Gell-Mann, Murray. *The Quark and the Jaguar*. New York: W. H. Freeman, 1994.

Herbert A. Simon
Carnegie-Mellon University, Department of Psychology, Pittsburgh, PA 15213

Near Decomposability and Complexity: How a Mind Resides in a Brain

The title of this book contains four very potent words: "mind," "brain," "complex" and "adaptive." The main task I have set for myself here is to consider the relation among the first three of these; I will have somewhat less to say about the fourth. Of course, it is generally agreed that the mind resides in the brain, and that mind and brain are complex (and adaptive) systems. Is there more than this to be said about the matter? Let me begin with complexity.

Shortly after World War II a great interest arose in something that was called "General Systems Theory." After a time, that interest faded, and for a simple and compelling reason: It proved to be very difficult, indeed, to find properties that were common to systems "in general," thereby to fill general systems theory with any theorems or other content. Control theory, which is more closely related to adaptivity than to complexity, split off and became a solid but separate domain within engineering. Communication theory, whose interest surely does focus on complexity, did flourish—also as a separate domain—and contributed as well to the foundations of computer science; but its ties to a general systems theory were lost, and the dream of creating such a theory has nearly died.

I mention these matters because we should take care lest our present undertaking suffer a similar fate. There is plenty that we can say about the brain and the mind, but is there anything we can say that will connect them in a meaningful way to concepts as general and abstract as "complexity" or "adaptivity"? Let me begin with some comments on complexity.

COMPLEXITY

I will not undertake a formal definition of complexity. For our purposes, we can regard a system as complex if it can be analyzed into many components having relatively many relations among them, so that the behavior of each component depends on the behavior of others. That definition embraces a very large number of the systems in nature in which we are interested, ranging from molecules and microbes (and smaller systems) to mammals, metropolises, and megapowers (and even larger systems).

HIERARCHY

It has been observed for a very long time that most, though not all, of the systems we regard as complex share a very important architectural feature: they are constructed in levels, with each level resting on the one below it. To change the metaphor, they can be represented as sets of boxes nesting within sets of boxes, nesting within..., often to a considerable number of repetitions. A familiar example is the molecule, composed of atoms which are composed of electrons and nuclei, which are composed of elementary particles, which are composed of quarks. Another example is the biological organism, which is composed of organs, which are composed of cells, which contain organelles, which are composed of molecules, and so on. A third example is a human society, which is composed of economic, social, and religious organizations, these, in turn of subgroups, down to the level of families. This social example is more complex than the others, since each individual may belong to a number of the larger subgroups—a family, a business firm, a church, and so on.

It has sometimes been suggested that this hierachical form of natural structures lies in the eye of the beholder—that we describe complex systems to ourselves and to others in terms of subsystems in order to simplify the description and make it comprehensible to the human mind. This may well be true. Examining a large computer program line by line, without any hint of larger components of structure that organize these lines, is seldom a meaningful exercise. The widely accepted doctrine of "structured" or "top-down" programming is aimed at using hierarchy in the program architecture to assure comprehensibility for programmer as well as user.

But hierarchy is not just a subjective matter, a convenient way of representing systems to aid human comprehension; it is an empirically detectable property of many natural systems, just as it is of programs that have been fashioned according to the prescriptions of structured programming. Moreover, it is no accident that most of the complex structures we encounter in nature have this hierarchical form. If we start with anything like a Big Bang, then large structures must emerge by means of some sort of combinatorial process from the small, relatively simple, stable structures that provide the initial raw material for the assembly process. It has been shown[13] that the probability of a large structure being formed in one step from a multitude of small ones is much lower than the probability of the assembly taking place in numerous successive stages, each one putting together a stable system from a few components that are a little smaller and simpler than the resulting product.

We all know that chemical reactions can usually be analyzed into a succession of steps, during each of which just a single association between two particles, or a single dissociation, takes place. If the reaction is more complex, uniting, say, three or more ions, a finer analysis with higher temporal resolution will usually show that the reaction can be elaborated into two or more simpler and successive reactions. The complex reaction is a highly improbable event compared with the succession of simpler reactions. For these reasons we can safely conclude that hierarchy— complex systems composed of a few less complex stable components, each of these of a few even simpler components, and so on—will be a dominant architectural form among natural systems. But hierarchical systems of this kind have an important special property that is not shared by systems with more amorphous structures: the property of near-decomposability.

HIERARCHY AND NEAR-DECOMPOSABILITY

Consider a dynamic system describable by N differential equations in N unknowns. We can represent this system by a matrix of the coefficients of the variables in the several equations. For simplicity of exposition, we will assume the equations to be linear and, consequently, the coefficients of the matrix to be constants. In a hierachical system, the variables in any particular subsystem will have, on average, relatively strong interactions with the other variables in that same subsystem but, on average, much weaker interactions with the variables belonging to other subsystems. If that is so, we can transpose rows and columns of the matrix so that columns corresponding to any given subsystem will be adjacent and rows corresponding to the equations describing the interactions within that subsystem will also be adjacent. Then, since all of the large coefficients, representing strong interactions, will be within these blocks, and only small coefficients outside them, the transposed matrix will be nearly block diagonal: all of the large coefficients will appear within one of a series of blocks along the main diagonal, while only small coefficients will appear in the remainder of the matrix.

If the matrix of a dynamic system is block diagonal, with only zero coefficients outside the blocks, then one can solve independently for the values of the set of variables within each block. If the matrix is nearly block diagonal, ignoring the coefficients outside the blocks and solving the resulting equations will still provide approximate solutions for the original matrix.[16] What this means from a dynamic standpoint is that if the variables in a block are disturbed from equilibrium, the equilibrium among them will be restored rapidly; over any longer span of time, each block will appear to move in a rigid fashion, the variables remaining close to equilibrium in relation to each other. Over some still longer period of time, the weaker interactions among different blocks will establish equilibrium of the entire system. The weaker the between-block interactions compared with the within-block interactions, the smaller will be the ratio of the time required to restore within-block equilibrium to that required to restore between-block equilibrium.

The argument can be extended to a hierarchy containing more than two layers. The blocks at the lowest level will equilibrate over the shortest time span; those at the next level, more slowly—with a restoration time that may be one or more orders of magnitude greater. The top level of the system will have the longest time for restoration of equilibrium. (The mechanism here is very close to the mechanism that underlies renormalization in physics.)

Visualize a building composed of many rooms, each room of many cubicles. Assume that there is a radical disequilibrium of temperature in the building, cubic foot by cubic foot. The temperature within each cubicle will equalize at some average value very rapidly—say, within a matter of minutes. The temperatures of the cubicles within each single room, separated by partitions but not by thick walls, will equalize a little more slowly—say, in a half hour. The temperatures of the various rooms will equalize still more slowly as heat passes through the walls that separate them. After some time, say, a few hours, all of the rooms of the building, all of the cubicles in each room, and each cubic foot of air within each cubicle will have the same temperature.

Near-decomposability is of great assistance in gaining an understanding of hierarchical complex systems. Suppose we have a system of N variables. To predict its behavior, we must solve N simultaneous equations. Suppose, however, that the system is composed of m stable subsystems each containing k variables, where $mk = N$. Then, we can predict the behavior of each subsystem by solving only k simultaneous equations. Having found the solution for each subsystem, we can replace the original system of equations by a simpler aggregate system (technically, the system described by the principal eigenvalues and principal eigenvectors of the subsystems). We then obtain the low-frequency response of the total system by solving only m simultaneous equations, one for each subsystem. Roughly speaking, the details of structure of the subsystems affect only their high-frequency behavior in returning to internal equilibrium; they do not affect the low-frequency behavior of the total system. With lower time resolution, the details of behavior of the subsystems do not "show through" to the behavior of the whole system. Only aggregate or average properties of the subsystems are important to that behavior.

HIERARCHY AND THE ORGANIZATION OF THE SCIENCES

The fact that most natural systems are hierarchical has had a major influence upon the division of science into its separate disciplines and upon the nature of the relations among these disciplines. Although molecules are assembled from atoms, atoms from electrons and nuclear particles, nuclear particles from elementary particles like quarks, we can have (and for a long time did have) some chemical theories of molecules without a theory of atoms, a theory of atoms without a theory of nuclear particles, and a theory of nuclear particles without a theory of elementary particles. Even though there are important relations between the successive levels of phenomena, we still have today the specializations of molecular biology and biochemistry, organic chemistry, physical chemistry, physics of continuous matter, atomic physics, nuclear physics, and the physics of elementary particles (which, itself, can be viewed in two or more layers). Although other factors are involved, this taxonomy of disciplines reflects, in large measure, the hierarchical structure of the phenomena that these disciplines, collectively, address.

Physical science and biological science have been built from the top down, with the aid of some sort of sky hook. Perhaps it would be more accurate to say, "from the middle up and down," for we have progressed gradually, since Dalton and Lavoisier, from the chemistry of simple molecules upward to biology and downward to quarks and, perhaps, strings. The claim that we can construct different sciences for different levels, each semi-independent of those above and below it, does not in any way imply holism or the irreducibility (at least in principle) of theory at one level to that of the next below. But the *detail* of behavior at each level does not much affect the behavior at the level above except in some average way, where the aggregates of interest often correspond to the principal igenvalues and eigenvectors of the system at the lower level. These parameters dominate and characterize behavior when the subsystem is near equilibrium and guarantee that in the longer run, the lower level system will behave "rigidly," that is, with few degrees of freedom.

There is one particular consequence of this state of affairs that deserves a brief mention because Roger Penrose has made much of the matter in a widely discussed book. Contrary to Penrose's view, it is unlikely that quantum uncertainty has any significant consequences for events at the level of organisms or their thought processes. Quantum uncertainty is an especially improbable candidate as the mechanism for free will, if there is such a thing, for the only freedom it supports is the freedom to behave randomly. It does not at all provide the kind of control of actions that would be required to make sense of the notion that a person "chooses" his or her behaviors.

Because of quantum uncertainty, certain macroscopic theories may have to remain irreducibly probabilistic (e.g., theories that depend upon genetic mutation), but averaging on the macroscopic level will smooth out the uncertainties, as they do in the transition from statistical mechanics to thermodynamics. Evolution, with its successive, hierarchy-producing processes, and layers of stable, rapidly equilibrating subsystems, has seen to it that microscopic phenomena are generally screened out from influence upon behavior at the macro level. The most important "leakage" from the molecular level to whole organisms is the evolutionary leakage from mutations and recombinations of genetic materials; and this occurs indirectly, through the high-level feedback mechanism of natural selection, and on a time scale of generations, not microseconds.

Thus, a few parameters (atomic weights, configuration of planetary electrons, etc.) characterize the atoms in molecular reactions; a few parameters of the nuclear particles (numbers of protons and neutrons) characterize the interactions of the nucleus with its planetary electrons, and so on down to the quarks. As we descend, higher and higher energies dominate, hence higher process speeds, returning the subsystems more and more rapidly to equilibrium. We can have a system of many levels with an essentially constant number of important degrees of freedom at each level. The complexity of a system at any given level depends largely on the steady states of the levels beneath it, and little on their dynamics.

I will not try to justify these statements here; the details are given in my earlier hierarchy paper, and a still earlier paper with Ando[16] describing nearly decomposable systems. I wish to discuss here the architecture of a particular system, the mind, and the characteristics of that architecture that make it possible to construct a powerful theory of human thinking, motivation, and emotion at the symbolic level, even in the absence of any but a very incomplete and primitive theory of how these symbolic processes are implemented by neuronal structurese.

Implemented they surely are—I repeat that I am not arguing for holism. But for the reasons I have just given, the theory of the mind's symbol structures and symbol processes is largely independent of the details of the neuronal implementation (but probably not independent of the more aggregate features of brain localization, as I shall note later). In the architecture of the sciences, there is a neuroscience of the brain at one level and a symbolic science of the mind at the level just above. Cognitive science is an umbrella term that embraces both of these (together with some other things).

As with most other complex systems in nature, the hierarchical structure of the mind reduces its operational complexity to that of a far smaller system, enhancing thereby the lawfulness and reliability of its operation, and preventing most "leakthrough" of minute disturbing events from its microscopic substructure.

THE PHYSICAL SYMBOL SYSTEM HYPOTHESIS

The explanation of thinking in terms of symbolic processes is summed up in the physical symbol system hypothesis. I will first say what I mean by "symbol," then describe the hypothesis.

PATTERNS AND SYMBOLS

A symbol is simply a pattern, made of any substance whatsoever, that is used to denote, or point to, some other symbol, or some object or relation among objects. The thing it points to is called its *meaning*. Words in our native language, printed on paper or formed as sound waves or as electromagnetic patterns, are familiar examples of symbols that denote objects, actions, or relations of all sorts. But not all symbols are linguistic in character: symbolic representation is not synonymous with representation in language. Diagrams are also symbols, having the special property that their structures are roughly homeomorphic with the structures of their meanings. The structures produced in the memory by the transduction of visual, auditory, haptic, or other stimulation are also symbols, denoting the stimuli that produced them. Diagrams and other nonlinguistic symbol structures can be used to make inferences just as sentences can; sometimes far more easily, sometimes with greater difficulty.[3]

I cannot emphasize too strongly that nonlinguistic symbols play at least as large a role in thought, and probably even a larger role, than do linguistic symbols. Any pattern whatsoever can become a symbol, simply by being used to point to or designate something. Patterns of retinal stimulation are symbols, for they designate the objects or scenes that were the source of the stimulation.

PHYSICAL SYMBOL SYSTEMS

A digital computer provides a contemporary example of a physical symbol system. The symbols are patterns of electromagnetism stored in memory, their physical nature varying with the age and model of the computer. Computers were originally invented to process patterns denoting numbers, but they are not at all limited to that use. The patterns stored in them can denote numbers, or words, or lizards, or thunderstorms, or the idea of justice. If you open a computer and look inside, you will not find numbers (or bits, for that matter); you will find patterns of electromagnetism.

Computers are capable of performing a few fundamental operations on symbols: They can (1) input symbols (e.g., from a keyboard or a sensor); (2) output symbols (e.g., print on paper, or initiate a physical action); (3) store symbols and relational structures of symbols in a memory; (4) construct, modify and erase such symbol structures; (5) compare two symbols structures for identity or difference;

and (6) follow one course of action or another, depending on the outcome of such a comparison. Matching symbol structures for identify or difference is the basis for the important process called "recognition."

The physical symbol system hypothesis[8] asserts that the necessary and sufficient condition for a system to be capable of thinking is that it be able to perform the symbolic processes that I have just enumerated. By thinking, I mean doing those things that, if they were done by a human being, we would call thinking—solving problems, reading, playing chess, making an omelette, or whatnot.

If the hypothesis is true, several consequences follow. First, computers, appropriately programmed, are capable of thinking. Second, the human brain, since it is capable of thinking, is (at least) a physical symbol system. These are empirical hypotheses, to be tested by spelling out some of their observable consequences and checking whether they are true. To test sufficiency, we write computer programs that enable computers to perform and to learn to perform the tasks that humans perform when they are thinking. Testing necessity is more difficulty: we must check whether the processes people use when they are thinking are just the symbol processes listed above.

Both parts of the symbol system hypothesis have, in fact, been tested extensively over the past 37 years, and the evidence has consistently supported it over a wider and wider range of human thinking tasks. [For samples of the evidence, see Newell and Simon[8] and Simon[15] (chapters 3 and 4).] Of course, it has not yet been conclusively demonstrated that computers can do *all* kinds of thinking of which humans are capable. Plenty of work remains in order to test the full scope of the hypothesis.

Whether the processes a computer uses in thinking are similar to the processes people use depends on how the computer is programmed. Consider, for example, computer programs that play chess. The most powerful of these, currently Deep Thought, plays at grandmaster level, but it demonstrably does not play in a humanoid way. It may examine as many as 50 million or more continuations to a position before it makes its move, while there is excellent experimental evidence that grandmasters seldom examine more than 100, and certainly never more than 1,000—numbers four or five orders of magnitude smaller than Deep Thought's numbers.

Deep Thought uses its computational power to "try everything" (of course, not really everything, but a great many possibilities). The grandmaster, with far weaker computational powers, uses chess knowledge to select the few lines of play that need to be examined. It is a contrast between nearly exhaustive search (at least to a moderate depth) and highly selective search.

On the other hand, there exist some chess programs less powerful than Deep Thought that have considerable skill, especially in sharp tactical positions, in selecting moves after searches no more extensive than the human ones. The PARADISE program, for example, can discover many deep mating combinations after exploration of less than 100 continuations.[19] PARADISE can serve as a first approximation to a theory of the symbolic process in human chess play; Deep Thought cannot.

MIND AND BODY

It long seemed mysterious that a material system, like the brain or like a computer, could have thoughts. Even dialectical materialists experienced great difficulty in accommodating mind in matter. To deal with the problem, Lenin introduced two kinds of matter: ordinary matter and "mind matter," and with the sanction of this dichotomy, doubts that computers can think have been held as widely and expressed as freely in Marxist societies as in the others.[17]

Part of the answer to the mind-body problem was clear as soon as humankind invented drawn pictures and written language. Drawings can denote; so can patterns on paper or parchment or stone. The old-fashioned IBM punched cards are quintessential examples of symbols.

As soon as we had computers capable of manipulating the patterns that they ingested and stored, we had found the other half of the answer: thinking meant processing these patterns in certain ways. Mind, Lenin's "mind matter," was manipulable pattern in tangible substances, whether neurons or transistors or chips. The mind-body problem ceased to be a problem. We know now how the functions of mind can be accomplished by a patterned body, a physical symbol system. Computers programmed to think successfully provide a constructive proof, making the possibility a reality.

SYMBOLS AND NEURONS

Computers and brains embody physical symbol systems having wholly dissimilar "hardware" implementations. Neurons and chips have little in common beyond the fact that they can both think, and that the latter can be programmed to think in humanoid fashion. This fact demonstrates the existence of a symbolic (software) level of theory above the hardware or neuronal level. The program of a computer is a set of difference equations that determine the action and next state of the computer as a function of its input and current state at any moment. In the same way, we can describe the patterns embodied in the brain's neural structures as a set of difference equations that determine human action and subsequent memory state as a function of sensory inputs and the current memory state.

Moreover, whenever a computer program can perform certain human thinking tasks in such a way as to track closely the sequence of behaviors in the human performance, that program (that set of difference equations) becomes a theory of the human thought processes in these task domains. This correspondence has been attained in many domains, the behavior of human and computer matching well within a temporal resolution of a few seconds. Many examples can be found in papers by Newell and Simon,[8] and by Simon.[14] Verbal thinking-aloud protocols and eye movements are the data most often used to match the human processes, step-by-step, with the computer trace.

Since the simplest symbolic processes (e.g., recognition of a familiar object) take at least a half second, the sorts of data we have available allow us to identify the processes being used in thinking pretty reliably down to the symbolic level. They do not give us many clues about the underlying neural events, whose durations are in the range from a millisecond (the time for a signal to cross a synapse) to a few hundred milliseconds. The limits of our methods of observation are regrettable, but still permit the construction and testing of symbol-level theory.[6] (Chapter 3).

One important kind of link is already being forged, however, between biological and symbolic theories of thinking. Insofar as the symbolic processes are localized, neurological evidence showing that some functions are performed in particular areas of the brain can help to support or refute proposed functional descriptions at the symbolic level. For example, recent work on the prefrontal cortex of monkeys shows it to be the locus of specialized working memories that encode identifying features of an object or of a spatial location, as the case may be.[2]

Currently, work on brain localization provides the principal bridge between theories at the symbolic and neural levels, and substantial converging support for seriality in higher-level processes. As we would expect from the discussion of levels in complex systems, this linkage between the biological and symbolic levels largely involves rather aggregated properties of the former (brain regions).

THE PROCESSES OF THOUGHT: SYMBOLIC LEVEL

The theory of thinking that has emerged from the research I have just described is reviewed by Simon[15] (chapters 3 and 4), by Newell and Simon,[8] and by Newell.[6] Two main processes predominate. The first, already illustrated by chess, is problem solving by selective search through a problem space, guided by heuristics, rules of thumb that suggest the more promising paths.

The second process is simply recognition of salient features in a situation that gives an experienced person reliable cues about how to respond to it. The human memory is organized like a very well indexed encyclopedia. Perceptual cues are the index items, which give access to stored information about the scene perceived and about relevant actions.

World class experts, in all of a wide range of occupations that have been studied, devote no less than ten years of intense application to acquiring these cues, or "chunks," and the information associated with them. It has been estimated that chess grandmasters, for example, have stored in memory no less than 50,000 chunks—patterns that they will notice and recognize if they appear on the board during a game, and that will give them access to the information needed to select a move.

These recognition capabilities account for experts' abilities to respond to many situations "intuitively," that is to say, very rapidly, and often without being able to explain why they responded as they did. Many, if not most, of experts' responses to routine, everyday situations are of this intuitive kind—based on recognition of familiar cues. We do not need to hypothesize additional mechanisms to explain intuition or insight.

The same recognition mechanism explains the "aha" that a scientist experiences in the face of a surprising phenomenon—Fleming's response to the bacteria in his dirty Petri dish being lysed by a mold, the Curies' surprise at the high level of radioactivity in their partially refined pitchblende, and many others. The knowledgeable and experienced scientist is surprised when events depart from expectations, and takes actions to exploit the surprise. Without prior knowledge and experience, there can be no expectations, hence no departure, hence no surprise.

In the course of thought of more than a second's duration, most of the inputs for mental operations must be present in a rapidly accessible short-term memory of limited capacity; and the outputs of the operations appear in that same memory. The thinker is generally aware of the contents of short-term memory, and can often report them orally while performing a mental task. Thus, when we add a column of figures, we are generally aware of each successive figure in the column and of the running total. Short-term memory is the principal locus of consciousness. As we shall see, the short-term memory is an extremely important feature of the brain's architecture (we have little knowledge of its biology), for it largely serializes the thought process.

NEUROLOGICAL THEORIES OF THINKING

In spite of the substantial recent progress in neuroscience, our clues as to how symbolic events are to be linked to neural events, are still largely limited to connections, referred to earlier, between hypothesized symbolic functions (e.g., the operation of working memory) and regions of the brain identified as performing these functions. We do not have a physiological theory of the "engram"—the basic mechanism of storage of information in the brain. Nor do we understand what biological mechanisms operate on this stored information. We know a great deal about how the retina works, and the pioneering work of Hubel and Wiesel gave glimpses of mechanisms for extracting features from retinal information transmitted to the brain. Although we have extensive knowledge at the symbolic level of the subsequent processes of recognition, and of reasoning upon information that has proceeded through the feature-recognition process, the detailed neurology of these processes lies largely in *terra incognita*.

A great deal is known about the biochemistry of the transmission of neural impulses. Apart from this, the major recent advances in neuroscience are mainly on a larger scale: advances in understanding the actions of some of the chemical substances that control the rate and effectiveness of brain functioning and the affect associated with it; and, as already noted, a growing knowledge of the localization of defined mental functions.[18] None of these advances have yet explicated the specific neural mechanisms that implement symbolic processes.

This state of affairs tells us that we must continue to pursue the research on thinking on both neural and symbolic fronts. There is plenty of opportunity for rapid progress, and every reason to multiply our efforts to build theories at both levels simultaneously. The parallel progress in nineteenth century chemistry and physics provides an encouraging precedent for this strategy. Even when additional links between neural and symbolic levels are forged, we will still want to characterize most of the higher-level and more complex cognitive phenomena at the symbolic level, rather than attempting to describe them solely in neurological terms. Aggregative descriptions of phenomena are not dispensable when we learn how to disaggregate them.

SERIAL OR PARALLEL?

Until quite recently, digital computers have been predominately serial machines. That is, although they have very large memories holding much information, the processes that act on this information are of a serial, one-at-a-time, kind. Computer programs exercise rather close control over the sequence in which these actions are performed. On the other hand, in the billions of neurons of the brain and the myriads of synaptic connections among them, there appears to be almost continuous widespread simultaneous activity. Some investigators (but not, by any means, all) have concluded from this that the brain is predominately a parallel device,[10] and sometimes they have further concluded that for this reason a serial computer cannot simulate brain processes.

This argument does not take levels into account. Even extensive parallelism at the neural level would not imply parallelism at the symbolic level. Since we do not know, from a neurological standpoint, how information is maintained in stable form in memory, much continuous and diffuse brain activity may be required simply for memory maintenance. Moreover, the presence of dispersed activity in the brain may mean that each event at the symbolic level requires a sequence of neural transmissions, which may not all be localized in any single small part of the brain. The temporal resolution of our observations of neural events is simply not great enough to choose among these and other interpretations of the widespread continuous activity that can be observed in the brain.

On the other hand, the information that we do have about brain localization gives substantial evidence of elaborate local specialization, and of sequential transmission of signals. It certainly does not support a picture of a highly undifferentiated and dispersed parallel network mechanism.

SERIALITY AT THE SYMBOLIC LEVEL

Important behavioral evidence can be brought to bear on the respective roles of serial and parallel processes in thinking at the symbolic level. We have already noted the central role of limited short-term memory capacity in serializing behavior. Humans are simply unable to carry on simultaneously more than one or two thinking tasks that are "hard." It is true that we can often carry on a conversation while driving a car; but we are probably time-sharing between the tasks rather than carrying them on simultaneously, for the conversation deteriorates as the traffic becomes heavier. (I speak with feeling and very direct knowledge on this point. My first traffic accident in forty years was caused recently by trying to read highway directional signs while entering the city of Mantova in Italy during rush hour.) Simultaneous performance is difficult for any set of tasks that are not highly automated by lengthy practice.

The most convincing case for seriality can be made for those processes that make use of short-term memory—the memory that allows us to carry a phone number, but not much more, from the phone book to the telephone on the other side of the room. But all processes that call for conscious awareness, hence virtually all complex thought processes, require short-term memory and are thereby constrained within the serial bottleneck that it imposes.

Seriality is not limited, however, to those processes that employ short-term memory to hold their inputs and outputs. Even automated activities, which are executed without conscious awareness and do not make demands for short-term memory capacity, are often executed serially because of the precedence constraints that force certain of their processes to be performed before others.

A THEORETICAL BASIS FOR SERIALITY

The facts of seriality in human thinking seem clear enough. We do not have at the present time, as far as I know, any kind of comprehensive theory to explain the circumstances under which we would encounter seriality and those under which we would encounter parallelism in the behavior of complex systems. However, the alluring prospect of designing highly parallel "supercomputers," and the effort devoted to such design over nearly forty years has cast a great deal of light on the matter.

The enemy of parallelism is the need, in performing tasks, for temporal priority of certain subtasks over others. For example, in multiplying two multi-digit numbers, several multiplations by individual digits in the multiplier must be performed before the partial products can be added. The multiplier and the adder cannot operate in parallel. Whenever there are many precedence constraints, it becomes exceedingly difficult to design a parallel architecture in which all or most of the parallel components can be kept busy simultaneously most of the time. Of course, if they cannot, the purpose of parallelism has been defeated, at a considerable hardware cost. Nature guided by natural selection would not build parallel architectures where priority constraints caused such inefficiency.

There has been a long history of failure (beginning, perhaps, with the ILLIAC IV) in designing highly parallel general-purpose computers, for the reason just set forth. Typically, general-purpose designs with thirty or forty processors operate only four or five times as rapidly as a single processor—a very expensive way to gain speed. On the other hand, there is a growing list of successes in designing parallel computers that are specialized to handle particular classes of tasks, for example, array computers or special-purpose chess machines like Hightech. In these special-purpose architectures, the parallel components are matched with components of the task that are relatively independent and, hence, unlikely to be left idle by unsatisfied precedence constraints. Computer scientists are just beginning to construct a body of theory that would explain the successes and the failures in general but accurate terms.

SERIALITY AND PARALLELISM IN HUMANS

What lessons does this experience have for a theory of the design of the thinking apparatus of humans or other mammals? What facts, other than those already cited, would help us test such a theory?

First, people do perform a number of complex activities in parallel: including breathing, circulating the blood through heart action, digesting food, and so on. These activities are performed "automatically," but most of them, notably the first two, place some reliance on nerve impulses. However, these are all internal activities that can be performed wherever the body happens to be at any moment.

Other complex tasks can only be performed by moving the entire body to a particular location, and the same location is seldom suitable for several tasks at the same time. Most such tasks also call upon the entire body, or at least several of its organs, simultaneously and in coordination. These facts greatly reduces the call for parallel computing capabilities for planning and controlling the activities associated with these acts. It would be uneconomic to provide substantial parallel capacity for them, which would be unused most of the time. We can conclude, therefore, that such capacity would not arise through evolutionary selection.

Even the sense organs have only limited parallel capabilities. The eye gathers most of its information through the fovea, with a radius of only about one degree of arc. Reading experiments show that in reading, for example, people acquire information fixating only one or two words at a time. The periphery of the eye does have a parallel function—alerting its owner to sudden motion that may require the interruption of attention to deal with immediate events. Similar comments can be made about the ear, whose parallelism seems to relate largely to ingesting a whole spectrum of sound frequencies at once. Studies of dichotic listening show, again, that attention focuses quite narrowly on only one stream of information, although there is some monitoring of unattended sounds to permit interruption. In summary, the sensory system appears to use parallelism mainly to permit interruption of the organism when real-time response to a new stimulus may be required.

Similarly, on the motor side, parallel capabilities generally facilitate cooperation of different parts of the body in performing a single task. It is difficult, usually requiring a lengthy period of learning, to generate motor signals permitting several unsynchronized tasks to be performed simultaneously. Indeed, simultaneous performance is generally achieved by introducing synchronization, even if not required by the nature of the task. (Chewing gum while playing the piano might provide interesting evidence on this point.)

Such behavioral evidence as we have, then, indicates that evolution has been quite thrifty, providing parallel neural capabilities only where they can be used efficiently by the organism in the kinds of activities it is capable of performing. Thinking about several independent topics at once does not seem to be one of these.

PARALLELISM AND NETWORK MODELS

At a more microscopic level, on the other hand, the retina of the eye is most certainly a parallel device, and signals are transmitted in parallel from different parts of the retina to the brain. At the motor end of the neural system we have also noted evidence of at least moderate parallelism of action—for example, in the finger movements of a skilled pianist or typist. The evidence for seriality weakens and that for parallelism grows somewhat as we move away from the central processes, and deal with elementary events of shorter duration. I am aware of no evidences, beyond those I have mentioned, of parallelism (as distinguished from time sharing) at temporal resolutions of a second or more.

During the past decade there has been a considerable research activity aimed at building so-called connectionist nets or neural networks to model human thinking within a parallel architecture.[10,12] This activity will be discussed extensively during the course of this seminar. Those efforts that undertake to introduce "neurons" possessing relatively realistic biological properties have been limited largely to simple, low-level structures containing a few neurons and, hence, cannot yet be linked in any clear way to models or phenomena at the symbolic level.

On the other hand the "neurons" of connectionist models are more numerous, but they have few of the properties of real neurons, and provide a foundation for only very abstract models of thought processes. These models are best viewed as another species of symbolic model (parallel instead of serial) where the symbols (patterns) are often distributed rather than highly localized. Moreover, it has not yet been demonstrated that connectionist models can account for any wide range of complex cognitive performances in the domains of problem solving, use of language or reasoning. Almost all of the tested computer simulations today of these kinds of human thinking are serial symbolic models.

There is certainly every reason to continue research on parallel models of brain activities along both of the paths just mentioned. But on the basis of what we know about the hierarchical structure of complex systems, and from what neurological evidence tells us about brain localization and the temporal sequencing of signals, it is improbable that such models will supersede models of the serial kind as explanations of behavior at the symbolic level, except possibly for mechanisms at the sensory/perceptual or motor peripheries of the system. A more promising goal for research with connectionist or neural net models is to use them to explain the basic mechanisms of memory and of the elementary information processes: inputting, outputting, storing, modifying, and comparing symbols. We will need this kind of a bridging theory in order to account for symbolic processes in neurological terms. The bridging will not remove the need for theories at a more aggregated level any more than nuclear physics has removed the need for theories of biochemistry.

ADAPTIVENESS

Time does not permit me to take up the topic of adaptiveness at any length. There has, however, been a good deal of attention to the implementation of learning within physical symbol systems, and some comparison of these learning systems with human learning. I will mention three examples of learning mechanisms that have been tested.

The first is the EPAM system, which was originally designed to simulate human performance in rote learning tasks (e.g., paired associate and serial anticipation learning) of the kinds that are so widely prevalant in the psychological literature. EPAM, which is currently being extended to an even wider range of tasks, has shown itself capable of simulating, quantitatively as well as qualitatively, a broad range of human learning phenomena.[1,9]

The "front end" of EPAM, a discrimination net capable of growth through learning, resides at the very boundary between perceptual processes that we would expect to be parallel in nature, and those we would expect to be serial. Therefore, special interest lies in the ability of EPAM to perform tasks that have also been studied with parallel architectures, for example, the letter perception task that McClelland and Rumelhart[4] simulated in their early *Psychological Review* papers. In that case[9] both architectures gave good agreement with the experimental data, and we were able to identify the features of the architectures that made this possible. There is a need for much more experimentation along this parallel/serial boundary, and, indeed, more experimentation in general that compares the performance of different architectures on the same task.

The Soar system exhibits another kind of learning: it is capable of remembering, "chunking," and streamlining sequences of actions that proved useful in performing a task, thereby speeding up performance the next time the task or one similar is performed again.[6] A distinct, but related, form of chunking, "data chunking," is performed by EPAM in the course of its discrimination learning. This chunking has been used to explain certain memory capabilities of experts which would otherwise appear to violate limits on the capacity of short-term memory.

A third learning mechanism, also closely related to the Soar mechanisms, is the adaptive production system. A production system is a system composed of if-then rules, an architecture widely used in cognitive simulation and in so-called Expert Systems. An adaptive production system is able, by examining worked-out examples in a task domain, to construct new productions automatically and to add them to its store.[5] This procedure has been shown to correspond closely to an important form of human learning, and has been used as the basis for the design of a high school algebra and geometry curriculum.[20]

Most research on parallel connectionist systems has focused on their learning capabilities, so there are many opportunities to compare the behaviors of serial and parallel systems in this important aspect of human cognition.

CONCLUSION

In this chapter, I have described what it means to explain human thought at the symbolic level, characterizing the brain as a physical symbol system. I have shown why we need several levels of cognitive theory: a neurological level to deal with events of less than a few hundred milliseconds duration, and a symbolic level to deal with the more aggregated slower events. The hierarchical structure of psychological theories resembles the hierarchical structure of theories in the other sciences because, in the brain as elsewhere, hierarchy has provided nature with powerful means for building reliable complex systems by assembling collections of simpler stable components. Neuroscience and symbolic cognitive science must collaborate in order to create a comprehensive, well-formed psychology.

I have sketched briefly some of the leading generalizations about thinking that have emerged at the symbolic level, especially problem solving by heuristic search and problem solving by recognition. These processes are strongly molded by the bottleneck of short-term memory which guarantees that most events with duration of a half second or more will occur serially rather than in parallel. Symbolic models that do parallel processing will probably have their main applications at the perceptual and motor peripheries of the neural system rather than in the more central portions that are involved in problem solving and other "higher" activities.

I hope that I have conveyed some of the excitement now present both in symbolic cognitive science and in neuroscience, which, with the help of new tools of investigation, are moving ahead rapidly at the present time.

And finally, I hope I have explained why a physical system, a brain, or a computer can exhibit the properties of mind. Understanding the human mind as a part of the system of nature is essential, for it becomes ever clearer that the survival of humankind depends on our ability to find for our species its appropriate role in that system, and its proper relation with the Earth and the other creatures that inhabit it.

ACKNOWLEDGMENTS

This research was supported by the National Science Foundation, Grant No. DBS-9121027; and by the Defense Advanced Research Projects Agency, Department of Defense, ARPA Order 3597, monitored by the Air Force Avionics Laboratory under contract F33615-81-K-1539. Reproduction in whole or in part is permitted for any purpose of the United States Government. Approved for public release; distribution unlimited.

REFERENCES

1. Feigenbaum, E. A., and H. A. Simon. "EPAM-Like Models of Recognition and Learning." *Cog. Sci.* **8** (1984): 305–336.
2. Goldman-Rakic, P. S. "Working Memory and the Mind." *Sci. Am.* **267(3)** (1992): 111–117.
3. Larkin, J. H., and H. A. Simon. "Why a Diagram is (Sometimes) Worth 10,000 Words." *Cog. Sci.* **11** (1987): 65–99.
4. McClelland, J. L., and D. E. Rumelhart. "An Interactive Activation Model of Context Effects in Letter Perception: Part 1. An Account of Basic Findings." *Psych. Rev.* **88** (1981): 375–407.
5. Neves, D. M. "A Computer Program that Learns Algebraic Procedures by Examining Examples and Working Problems in a Textbook." In *Proceedings of the Second National Conference of the Canadian Society for Computational Studies of Intelligence*, 1978.
6. Newell, A. *Unified Theories of Cognition*. Cambridge, MA: Harvard University Press, 1990.
7. Newell, A., and H. A. Simon. "Computer Science as Empirical Inquiry: Symbols and Search." *Comm. ACM* **19** (1976): 113–126.

8. Richman, H. B., and H. A. Simon. "Context Effects in Letter Perception: Comparison of Two Theories." *Psych. Rev.* **96** (1989): 417–432.

9. Rumelhart, D. E. "The Architecture of Mind: A Connectionist Approach." In *Foundations of Cognitive Science*, edited by M. I. Posner. Cambridge, MA: MIT Press, 1989.

10. Rumelhart, D. E., and J. L. McClelland. "An Interactive Activation Model of Context Effects in Letter Perception: Part 2. The Contextual Enhancement Effect and Some Texts and Extensions of the Model." *Psych. Rev.* **89** (1982): 60–94.

11. Sejnowski, T. J., and P. S. Churchland. "Brain and Cognition." In *Foundations of Cognitive Science*, edited by M.I. Posner. Cambridge MA: MIT Press, 1989.

12. Simon, H. A. "The Architecture of Complexity." *Proc. Am. Phil. Soc.* **106** (1962): 467–482.

13. Simon, H. A. *Models of Thought*, Vols. 1 and 2. New Haven, CT: Yale University Press, 1979.

14. Simon, H. A. *The Sciences of the Artificial.* Cambridge MA: MIT Press, 1981.

15. Simon, H. A., and A. Ando. "Aggregation of Variables in Dynamic Systems." *Econometrica* **29** (1961): 111–138.

16. Simon, H. A., and Q. Jing. "Raspoznavanie, mishlenie i obychenie kak informatsionnie processi." *Psikhologicheskii Zhurnal* **9** (1988): 33–46.

17. Thompson, R. F., N. H. Donegan, and D. G. Lavond. "The Psychobiology of Learning and Memory." In *Stevens' Handbook of Experimental Psychology*, edited by R. C. Atkinson, R. J. Herrnstein, G. Lindzey, and R. D. Luce, Vol. 2, 245–347, 1988.

18. Wilkins, D. E. "Using Knowledge to Control Tree Searching." *Art. Intl.* **18** (1982): 1–5.

19. Zhu, X., and H. A. Simon. "Learning Mathematics From Examples and by Doing." *Cog. & Inst.* **4** (1987): 137–166.

John H. Holland
Division of Computer Science and Engineering, University of Michigan, Ann Arbor, MI 48109

Can There Be a Unified Theory of Complex Adaptive Systems?

Many of our most troubling long-range problems—trade balances, sustainability, AIDS, genetic defects, mental health, computer viruses—center on certain systems of extraordinary complexity. The systems that host these problems—economies, ecologies, immune systems, embroyos, nervous systems, computer networks—appear to be as diverse as the problems. Despite appearances, the systems have enough significant characteristics in common to make it possible, even probable, that common general principles explain their dynamics. For this reason, we group these systems under a single classification at the Santa Fe Institute, calling them complex adaptive systems (CAS). This is more than terminology. It signals our intuition that there are general principles that govern all CAS behavior, principles that point to ways of solving the attendant problems. Much of our work is aimed at turning this intuition into fact.

It is an easy exercise to produce a list of significant characteristics common to all CAS: (1) All CAS consist of large numbers of components, *agents*, that incessantly interact with each other. (2) It is the concerted behavior of these agents, the *aggregate behavior*, that we must understand, be it an economy's aggregate productivity, or the immune system's aggregate ability to distinguish antigen from self. (3) The interactions that generate this aggregate behavior are nonlinear, so that the aggregate behavior cannot not be derived by simply summing up the behaviors of isolated agents. (4) The agents in CAS are not only numerous, but also diverse. An ecosystem can contain millions of species melded into a complex web of interactions; the mammalian brain consists of a panoply of neuron morphologies organized into an elaborate hierarchy of modules and interconnections; and so on. (5) The diversity of CAS agents is not just a kaleidoscope of accidental patterns; remove one of the agent types and the system reorganizes itself with a cascade of changes, usually "filling in the hole" in the process. (6) The diversity evolves, with new niches for interaction emerging, and new kinds of agents filling them. As a result, the aggregate behavior, instead of settling down, exhibits a perpetual novelty, an aspect that bodes ill for standard mathematical approaches. (7) CAS agents employ *interal models* to direct their behavior, an almost diagnostic character. An internal model can be thought of, roughly, as a set of rules that enables an agent to anticipate the consequences of its actions. Even an agent as simple as a bacterium employs an "unconscious" internal model when it swims up a glucose gradient in the search for food, while humans make continual prosaic use of internal models, as in our unconscious expectation that room walls are unmoving structures.

The combination of internal models with a diversity of agents, along with the attendant nonlinearities, undercuts most traditional approaches to system dynamics. Anticipations based on internal models, even when they are incorrect, may substantially alter the aggregate behavior. And the evolving diversity of agents in a CAS produces a perpetual novelty in dynamics. CAS will certainly remain mysterious until we can take such effects into account.

ADAPTIVE AGENTS

It is much easier to produce a definition of *adaptive agent* than it is to produce a general formal definition of CAS. The definition turns on the computation-based implementation of two processes[6]: (1) a performance system that specifies the agent's capabilities at a given point in time—its abilities in the absence of further learning—and (2) an inductive apparatus that modifies the performance system as experience accumulates—the learning mill.

The performance system is conveniently defined in terms of rules. Any given environmental situation is defined by the concurrent activation of a cluster of rules that describe, and act upon, parts of the overall situation. The rules are treated as hypotheses, rather than as facts, and are subject to progressive action, or replacement, as the system accumulates experience.

Because CAS consist of diverse interacting agents, the attainment of goals in this perpetually novel environment depends upon each agent's ability to distinguish among other agents in the system. To provide for this possibility, it is necessary that individual agents exhibit labels, banners, phenotypic identifiers, or the like. I use the collective term *tag* to designate such external markers.

Three mechanisms help the agent to balance exploration (acquisition of new information and capabilities) against exploitation (the efficient use of information and capabilities already available):

1. *Parallel execution* of rules allows transfer of experience to novel situations by the combined activation of relevant rules—building-block rules—that describe aspects of the situation.
2. *Tags and rule coupling* provide for directed sequential action, making a "bridge" for credit assignment to stage-setting actions. Tags also provide for adaptive clustering of rules.
3. *Competition*, based on rule specificity and strength, allows rules to be treated as hypotheses to be marshalled and progressively confirmed (by strength revision under credit assignment) as required by the changing environmental situation.

To provide new hypotheses, a genetic algorithm[5] treats strong rules as parents, recombining parts of the parents to provide offspring rules that replace weak rules (hypotheses). The resulting adaptive agent is well defined in computational terms. Via these mechanisms, it constructs increasingly sophisticated internal models (default hierarchies, for example) that enable it to anticipate its environment. Its predictions are continually tested against outcomes, with falsifications being used to improve the models, even in the absence of payoff. Though the agent readily improves its performance, it uses only computationally simple procedures to do so. As such, it conforms to reasonable notions of bounded rationality.

TAGS RECONSIDERED

Some experiments comparing the evolution of adaptive agents with tags to those without have been carried out.[10,12] The most recent experiments use the game matrix of the Prisoner's Dilemma (PD) to describe the outcome of individual interactions (Axelrod[1] provides a good introduction to studies of PD using genetic algorithms).

When agents without tags make repeated plays of the PD with random unidentified opponents, there is no useful basis for implementing conditional interactions, and the productive "tit-for-tat" strategy never evolves; interactions settle on the minimax defect-defect mode. Agents with tags evolve along an entirely different path. At some point, as the strategies evolve, an agent appears that (1) employs tit-for-tat and (2) has a conditional interaction rule based on a tag carried by a subpopulation that is susceptible to tit-for-tat. Such agents achieve a higher reproduction rate because of the higher rate of payoff under the cooperate-cooperate mode, causing both this agent and its cooperating partners to spread through the population. Subsequent recombinations provide agents that play tit-for-tat and restrict their interactions to other agents playing tit-for-tat. Once established, such a subpopulation is very resistant to invasion by other strategies (see Maynard Smith's[9] study of Evolutionarily Stable Strategies (ESS)).

Even in the limited confines of the population-based PD, the evolutionary opportunities for adaptive agents with tags go considerably beyond the ESS just discussed. For example, mimicry becomes possible. An agent can present the tag associated with tit-for-tat, while pursuing a different strategy. Thus, the presence of an agent with a tag that has a well-defined functional meaning—the tag "means" tit-for-tat in this case—opens new niches for other agents. It is interesting that these niches are usually constrained in size, depending as they do on the continued presence of the "founding" agent. The mimic, as biological studies suggest, can only occupy a small proportion of the overall population, relative to the agents being mimicked, because other agents begin to adjust to the deception as the proportion of mimics increases. This negative feedback sets a limit on the mimic's expansion. It is typical that tags provide niches of limited "carrying capacity," leading to highly diverse systems with no "super individual" that outcompetes all comers.

A BROADER PERSPECTIVE

Even a cursory look uncovers many examples of natural agents wherein tags encourage diversity and complexity. The studies of chemotaxis in the slime mold initiated by Bonner[2,8] and the studies of cell adhesion molecules initiated by Edelman[3,11] provide two sets of examples of chemical messages and receptors interacting to provide sophisticated organizations. Selective mating based on phenotypic characteristics,[4] and complex interactions in social insects mediated by pheromones[7] provide examples of tag-mediated interactions at a more macroscopic level. If these examples are typical, and I think they are, tags play a critical role in the phylogeny of CAS.

To go a step further, I think that tags are the central mechanisms that enable CAS to evolve a diverse array of agents that interact in an integrated way. Even a small ecosystem can involve hundreds of thousands of distinct species, a mammalian immune system involves like numbers of antibodies, and a municipality involves thousands of distinct kinds of social interactions. The diversity arises as tags and interactions evolve, yielding a complex web of interdependent interactions wherein the persistence of any agent depends directly on the context provided by the other agents. In the population-based PD, tags made diversity possible by breaking the unexploitable symmetry of random pairing. This observation can be elevated, I think, to a general principle governing CAS: Tags break symmetries, providing opportunities for diversity.

In order to study these effects, I have defined the ECHO class of models. The prototype of ECHO is a simulated closed world with almost trivial representations of geography, physics, chemistry, biochemistry, etc. ECHO does provide for a distinction between genotype and phenotype, so that the fitness of a genotype depends upon interactions of the phenotype with other agents and the local environment. Despite ECHO's simplicity and its completely endogenous character (it receives no inputs or control signals once it is started), this "world" exhibits counterparts of sophisticated ecological processes, such as biological arms races and speciation. More advanced versions of ECHO, involving agents displaying tags and involving simple grammars controlling the relation between genotype and phenotype, should exhibit counterparts of the generation and exploitation of niches (by parasitism, symbiosis, mimicry, etc.), selective mating and recombination, the evolution and spread of multifunctional coadaptive sets of alleles, and, most importantly, the counterparts of mechanisms (such as competence and induction in morphogenesis) that permit the evolution of sophisticated organizations.

On the basis of this perspective, I do believe a useful unified theory is possible. It would, I think, involve the following elements:

1. Interdisciplinary comparison. Different CAS show different characteristics of the class to advantage.
2. A "Correspondence Principle." Using Bohr's principle, *mutatis mutandis*, CAS should encompass standard models from prior studies of particular examples of CAS (such as the Prisoner's Dilemma, Wicksell's Triangle, Overlapping Generations Models, Lotka-Volterra models, and the like).
3. Computer-based gedanken experiments. Computer-based experiments, by varying parameters under different "restarts" from known initial conditions, allow a systematic search for invariants and critical patterns. Such experiments provide existence proofs of the sufficiency of given mechanisms for generating observed CAS phenomena.
4. A mathematics of competition-based recombination. A mathematics so-oriented would emphasize invariant features of evolutionary, far-from-equilibrium, trajectories generated by recombination, such as time to first occurrence of certain kinds of "building blocks," rate of spread of such "building blocks," mechanisms that maintain diversity, and so on.

REFERENCES

1. Axelrod, R. "The Evolution of Strategies in the Iterated Prisoner's Dilemma." In *Genetic Algorithms and Simulated Annealing*, edited by L. D. Davis, 32–41. San Mateo, CA: Morgan Kaufmann, 1987.
2. Bonner, J. T. "Evidence for the Formation of Aggregates by Chemotaxis in the Development of the Slime Mold *Dictyostelium discoideum*." *J. Exp. Zool.* **106** (1947): 1–26.
3. Edelman, G. M. *Topobiology*. Basic Books, 1988.
4. Hamilton, W. D. "The Genetical Evolution of Social Behavior." *J. Theor. Biol.* **7** (1964): 1–52.

5. Holland, J. H. *Adaptation in Natural and Artificial Systems*, 2nd ed. Cambridge, MA: MIT Press, 1992.

6. Holland, J. H., K. J. Holyoak, R. E. Nisbett, and P. R. Thagard. *Induction: Processes of Interference, Learning and Discovery.* Cambridge, MA: MIT Press, 1989.

7. Hölldobler, B., and E. O. Wilson. *The Ants.* Cambridge, MA: Harvard University Press, 1990.

8. Kessin, R. H., and M. M. Van Lookeren Campagne. "The Development of Social Amoeba." *Am. Sci.* **80(6)** (1992): 556–565.

9. Maynard Smith, J. *The Evolution of Sex.* Cambridge: Cambridge University Press, 1978.

10. Perry, Z. A. "Experimental Studies of Speciation in Ecological Niche Theory Using Genetic Algorithms." Ph.D. Dissertation, University of Michigan, 1984.

11. Sharon, N., and H. Lis. "Carbohydrates in Cell Recognition." *Sci. Am.* **268(1)** (1993): 82–89.

12. Riolo, R. Personal communication, 1992. Dr. Riolo is a post-doctoral researcher at the University of Michigan.

Patricia S. Goldman-Rakic
Yale University School of Medicine, 333 Cedar Street, New Haven, CT 06510

Neurobiology of Mental Representation

Psychologists have traditionally denied the relevance of brain to mind while neurobiologists have essentially ignored the mind, treating it as intractable and by default irrelevant. As the symposium indicated, the mind/brain discussion is now the subject of cross-disciplinary research with the result that mental phenomena are becoming recognized by brain research, and neurology and neurobiology are being increasingly acknowledged by behaviorists and theorists. Nevertheless, scholars and scientists on both sides of the debate remain skeptical of one another's approach and doubts prevail concerning the issue of whether neurobiology can add insight to cognition and mental processing and, conversely, whether cognitive sciences can enrich an understanding of brain function. In this chapter I hope to illustrate from recent research on nonhuman primates (1) that a genuine neurobiology of mental representation is possible; and (2) that significant principles concerning the organization of the human thought process can be derived from neurobiology. The work I will describe relies on behavioral analysis equally as much as on neurobiology.

PREFRONTAL CORTEX: THE CORTEX OF COGNITION

The functions of prefrontal cortex have been the subject of intense scientific curiosity and considerable speculation for most of this century and understanding its role in executive behavior can be considered one of the most significant challenges of modern neurobiology. Elsewhere we have presented evidence that the prefrontal cortex is specialized for a particular memorial process—working or representational memory.[10] Although mnemonic functions are considered important in every theory of prefrontal cortex, most views of prefrontal cortex have considered its memory function as only one of a multifaceted set of cognitive functions, including attentional functions, preparatory motor set, and interference control.[4,8,14,15,17,28] Indeed, over most of the present century, with few exceptions, the prefrontal cortex has been viewed as a structure with multiple functions separately localized in orbital and dorsolateral cortex or in subdivisions thereof and mnemonic processing has not been considered particularly central.

An entirely different view with respect to the role of memory processes in prefrontal function is that the prefrontal cortex has a global role in behavior "beyond that of any specialized function" through an involvement in *all types* of memory.[9] Gaffan describes several tasks on which performance is altered by prefrontal lesions in experimental animals but does not specify the nature of the different memory processes, nor if and how they are allocated among different areas of the prefrontal cortex. Gaffan appears to believe that different areas of the brain are dedicated to different mnemonic functions, and that the prefrontal cortex is engaged in all of these. I would point out only that lesions of the major subdivisions of prefrontal cortex—the dorsolateral and orbital prefrontal areas—fail to produce lasting impairments on many basic learning and memory tasks.[10]

I have advanced a still different view of prefrontal function which contrasts significantly with both of the positions so far mentioned mainly with respect to the central role and replicative organization of memory in the cognitive functions of the prefrontal cortex. I believe, as do many theorists, that the prefrontal cortex plays a global role in behavior and also agree that motor set and behavior control are affected by prefrontal damage. Others have associated working memory deficits with prefrontal cortex damage in humans.[29] However, I have gone further to advance the ideas (i) that working memory—the basic ability to keep track of and update information at the moment—is the cardinal specialization of the granular prefrontal cortex, (ii) that prefrontal cortex contains multiple segregated special-purpose working memory domains rather than one central executive center, and (iii) that both the cognitive capacities and cognitive deficits of humans can be derived from consideration of working memory as the unitary operation.[10] This process-oriented view does not deny the effects of prefrontal lesions on preparatory set or interference control on complex tasks but explains these effects as a default in the working memory system with input (sensory), delay (mnemonic), and response (motor) processing components.[11] This unified view holds that prefrontal cortex has a specialized function that is replicated in many, if not all, of its various cortical subdivisions and that the interactions and coactivation of these working memory centers within cortical networks together constitute the brain's machinery for higher-level cognition. In my opinion, this view is supported by the large experimental, clinical, and neurobiological database (summarized by Goldman-Rakic[10]).

MEMORY DICHOTOMIES: WORKING VS. ASSOCIATIVE MEMORY

Working memory is a concept developed by cognitive psychologists to refer to a distinct operation required for cognition, namely, the ability to update and/or bring information to mind from long-term memory and/or to integrate incoming information for the purpose of making an informed decision, judgment, or response.[2,13,19] As explained by Baddeley, a transient and active memory system to a large extent evolved from the older seminal abstraction, "short-term" memory, to explain the dynamic features of human memory.[1] Working memory can be distinguished operationally from canonical (associative) memory by several formal criteria: short duration and limited capacity, functional purpose, and neural mechanism. With respect to the latter, I have reviewed the considerable evidence from studies of prefrontal lesions in nonhuman primates showing that the performance of monkeys with prefrontal lesions is selectively impaired on tasks with working memory components while their performance is spared on the host of associative memory tasks that have been constructed to assess memory in experimental animals.[10] The delayed-response tasks are exactly the type of task that taxes an animal's ability to hold information "in mind" for a short period of time because the task demands that the memory be updated from trial to trial. It is important to note that all aspects of task performance that rely on associative memory—i.e., waiting for a signal to respond, executing a particular response on every trial, in brief, familiarity with the rules of the task and the motor requirements—do not cause a problem for the prefrontally lesioned animal. Its difficulty lies in remembering the specifics or content of the information that is needed to guide a correct response. Thus, I have suggested that the brain obeys the distinction between working and associative memory, and that prefrontal cortex is preeminently involved in the former; other areas such as the hippocampal formation and posterior sensory association regions, the latter.[10]

WORKING MEMORY: STORAGE AND PROCESS

It has been emphasized by cognitive theorists that working memory has at least two components—a storage component and a processing component.[1,13] The question can be raised as to whether working memory is sufficiently developed in nonhuman species and whether it can be studied in them. This question seems easily answered in the affirmative because it can be demonstrated that monkeys are capable of remembering briefly presented information over delays in a delay-dependent manner—in the classical spatial delayed-response tasks (for review see Fuster[8] and Goldman-Rakic[10]) as well as in the more demanding eight-item oculomotor version of that task[5] in various match-to-sample paradigms[21] and in "self-ordering" tasks.[23] It is less easily shown that monkeys can process information, i.e., transform it mentally. Shintaro Funahashi and Matthew Chafee addressed this issue in my laboratory by training monkeys on an anti-saccade task, similar to that used by Guitton et al.[12] to study the effects of unilateral frontal cortical damage in humans. The anti-saccade paradigm required the monkeys to suppress the automatic or prepotent tendency to respond in the direction of a remembered cue and instead respond in the opposite direction, a transformation that is not particularly easy for human subjects. In addition, we recorded from the prefrontal cortex in our trained monkeys to isolate and characterize neuronal activity in the principal sulcus and surrounding cortex. We implemented a compound delayed-response task in which, on some trials, the monkey learned to make deferred eye movements to the same direction signaled by a brief visual cue (standard oculomotor delayed-response [ODR] task) and, on other trials, to suppress that response and direct its gaze to the opposite direction (anti-saccade task, AS-ODR). The monkeys succeeded in learning this difficult task at high (85%, and above) levels of accuracy. In itself, their acceptable learning performance indicates that monkeys are capable of holding "in mind" two sequentially presented items of information—the color of the fixation point and the location of a spatial cue and of transforming the direction of response from left to right (or the reverse) based on a mental synthesis of that information. If "if-then" mental manipulations can be equated with propositional thought, the anti-saccade task may be a way of assaying the thought process in nonhuman species. Further, the task provides an elegant way of dissociating the direction of the cue from the direction of the response to allow us to determine the coding strategy of prefrontal neurons.

A major finding from these studies is that the great majority (approximately 60%) of prefrontal neurons was selectively activated during a silent three-second period intervening between a particular antecedent stimulus and the prospective response, *whether* the intended movement was toward or away from the designated target. In Figure 1 I illustrate a neuron that exhibited enhanced activity in the delay period of the ODR task whenever the visual cue to be remembered was presented on the right (Figure 1(a), left). For targets presented on the left, by contrast, delay-period activity was not above baseline (Figure 1(a), right). On the AS-ODR trials, the neuron was again activated preferentially in the delay period when the visual cue was presented in the right, in spite of the fact that the monkey's response was now directed to the left at the end of the delay. The absence of motor planning activity in this neuron is further demonstrated by the absence of

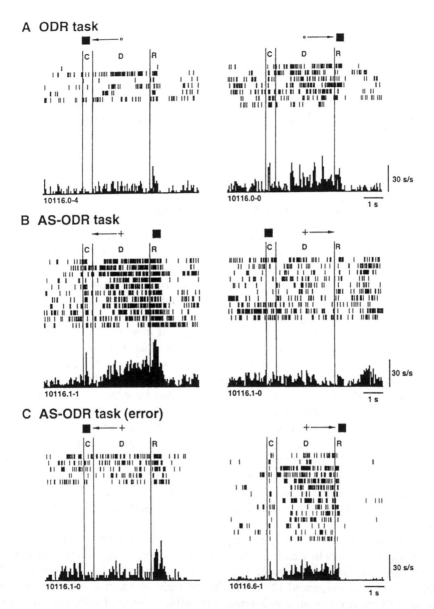

FIGURE 1   The neuron shown in the figure was tested during concurrent administration of ODR and AS-ODR trials. On ODR trials, the cell's activity was significantly higher during the delay when the target to be recalled was to the right compared to the left. The same pattern of firing obtained on the anti-saccade trials, even though the target on the right now signaled the animal to respond to the left. Thus, the neuron's activity was not dictated by the direction of responding but by the direction of the remembered cue. Modified from Funahashi, Chafee, and Goldman-Rakic.[7]

enhanced activity before saccades to the right in the anti-saccade task condition (Figure 1(b), right), demonstrating that it was not the rightward saccade in the ODR condition that was driving the unit. This result thus establishes that the same neuron involved in commanding an oculomotor response is also engaged when this response is suppressed and/or redirected. Such findings argue for at least a rudimentary form of propositional thinking on the part of nonhuman primates as well as point toward a cellular basis for mental processing in the nonhuman primate prefrontal cortex.

MULTIPLE WORKING MEMORY DOMAINS

According to the working memory analysis of prefrontal function, a working memory function should be demonstrable in more than one area of the prefrontal cortex and in more than one knowledge domain. Thus, different areas within prefrontal cortex will share in a common process—working memory; however, each will process different types of information. Thus, informational domain, not process, will be mapped across prefrontal cortex. Evidence on this point has recently been obtained in our laboratory from studies of nonspatial memory systems in prefrontal cortex.[20,33,34] In particular, we explored the hypothesis that the inferior convexity of the prefrontal cortex may contain specialized circuits for recalling the attributes of stimuli and holding them in short-term memory—thus processing nonspatial information in a manner analogous to the mechanism by which the principal sulcus mediates memory of spatial information. The inferior convexity cortex lying below and adjacent to the principal sulcus is a likely candidate for processing nonspatial—color and form—information, in that lesions of this area produce deficits on tasks requiring memory for the color or patterns of stimuli[18,22] and the receptive fields of the neurons in the posterior portion of this area, unlike those in the dorsolateral cortex above, represent the fovea, the region of the retina specialized for the analysis of fine detail and color—stimulus attributes important for the recognition of objects.[16,32]

We recorded from the inferior convexity region in monkeys trained to perform delayed-response tasks in which spatial or feature memoranda had to be recalled on independent, randomly interwoven trials. For the spatial delayed-response (SDR) trials, stimuli were presented 13° to the left or right of fixation while the monkeys gazed at a fixation point on a video monitor. After a delay of 2500 ms, the fixation point disappeared, instructing the animal to direct its gaze to the location where the stimulus appeared before the delay. For the pattern delayed-response (PDR) trials, various patterns were presented in the center of the screen; one stimulus indicated that a left-directed and the other a right-directed response would be rewarded at the end of the delay. Thus, both spatial and feature trials required exactly the same eye movements at the end of the delay; but differed in the nature of the mnemonic representation that guided those responses.

We found that neurons were responsive to events in both delayed-response tasks. However, a given neuron was generally responsive to the spatial aspects or the feature aspects and not both. Thus, a large majority of the neurons examined in both tasks were active in the delay period when the monkey was recalling a stimulus pattern which required a 13° response to the right *or* left. The same neurons did not respond above baseline during the delay preceding an identical rightward or leftward response on the PDR trials. Neurons exhibiting selective neuronal activity for patterned memoranda were almost exclusively found in or around area 12 on the inferior convexity of the prefrontal cortex, beneath the principal sulcus while neurons that responded selectively in the SDR were rarely observed in this region, appearing instead in the dorsolateral cortical regions where spatial processing has been localized in our previous studies. In addition, we discovered that the neurons in the inferior convexity were highly responsive to configurational stimuli, such as faces or specific objects. We subsequently used face stimuli as

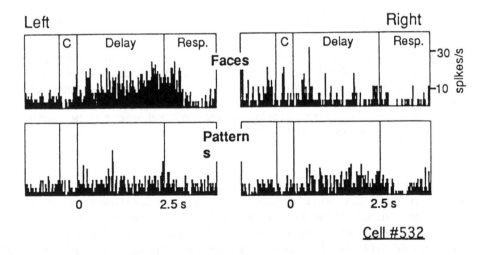

FIGURE 2 This neuron was activated in the delay when the stimulus to be recalled was a particular face (a: left panel); but not for another face (b: right panel). The same neuron was not differentially activated by the recall of patterned cues (c & d: lower panels). This result illustrates that prefrontal neurons can code selective aspects of or selected images in working memory. Arrows indicate direction of response for a given memorandum. Modified from Wilson, O Scalaidhe, and Goldman-Rakic.[34]

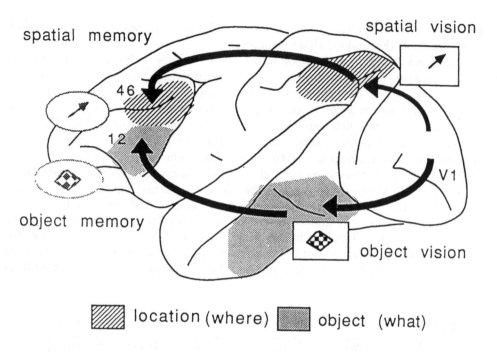

FIGURE 3 Multiple memory domains are illustrated on this diagram of the monkey prefrontal cortex. The dorsolateral area around the principal sulcus and anterior arcuate is important for spatial working memory, and that for features or attributes of objects, in the inferior convexity of the prefrontal cortex. The two areas are reciprocally interconnected with the posterior parietal and inferotemporal regions of the hemisphere, respectively. Modified from Wilson et al.[34]

memoranda in a working memory task and demonstrated that such stimuli could indeed serve as memoranda in memory tasks. In Figure 2 I show a neuron that encoded a face stimulus in the delay period of our working memory paradigm (Figure 2(a)). The same cell was unresponsive on trials when the monkey had to remember a different face (Figure 2(b)) as well as when patterns were used as memoranda (Figures 2(c)–2(d)). It should be noted that even though the same response is required on trials shown in Figure 2(a) and 2(c), the neuron responds in the delay only in Figure 2(a). These results provide strong evidence that the neuron in question is encoding information about the features of a stimulus and not about the direction of an impending response. Altogether results establish that nonspatial aspects of an object or stimulus may be processed separately from those dedicated to the analysis of spatial location and vice versa. Furthermore, different features are encoded by different neurons. Thus, feature and spatial memory—what and where an object is—are dissociable at both the areal level and the single neuron level. Altogether these findings support the prediction that different prefrontal subdivisions represent different informational domains rather than different processes. Thus, more than one working memory domain exists in prefrontal cortex (see Figure 3).

IMPLICATION FOR HUMAN COGNITION: ACTIVATION STUDIES USING PET

The application of positron emission tomography and other promising methods like fast-scan magnetic resonance spectroscopy for the study of cognitive activation in human subjects offers an unprecedented opportunity to test hypotheses about cortical organization derived from studies of experimental animals and from human neuropsychology. The prefrontal cortex, in particular, is activated by working memory tasks and several recent studies have pinpointed areas 9 and 46, in particular, as functional sites in the normal performance of various memory processes. Petrides et al. have recently provided evidence that area 46 was activated in both verbal[25] and nonverbal[26] working memory tasks. Likewise, Frith et al.[6] have reported that area 46 was activated during tasks calling for the open-ended generation of words or finger movements. All of the tasks employed in these studies had a working memory component in that the information needed to guide correct responses was not present in the immediate environment but had to be retained in working memory and/or generated de novo at the time of response. Importantly, control tasks were used in both studies that required similar or identical motor responses but these responses were either repetitive (selecting the same stimulus over and over again), routine (alternating finger taps), or instructed by external cues (if red, then apple)—conditions that depend upon associative memory processes. None of these control conditions activated the dorsolateral cortex, Brodmann's area 46.

The idea that different cortical subdivisions constitute working memory centers for different information domains also obtains some support from these studies of human cognition. Different cortical areas could be expected to be activated when verbal working memory is required as compared with solving largely nonverbal tasks that require self-referent orientation such as the self-ordering task used by Petrides et al. and a finger-tapping task used by Frith et al. The documentation of Tailarach coordinates in each of these studies allows tentative determination of whether the same cortical sites were activated across studies and/or whether the same portions of the designated areas were activated. In addition, we have been able to reanalyze these coordinates in relation to a cytoarchitectonic remapping of areas 46 and 9 in the human brain carried out in our own laboratory.[27] Our analysis of the data presented by Frith and Petrides supports their conclusions that area 46 was activated by the self-ordering task[25,26] and the finger-tapping task[6] but does not support their belief that their verbal tasks engaged the same area. Our analysis suggest that the verbal tasks of both studies may activate area 45 and/or area 44. If our analysis is correct, it reveals that cognitive functions in humans are segregated by domain and that the human prefrontal cortex may contain multiple working memory centers, as does the monkey. This conclusion would be more in keeping with the vast anatomical, clinical, and experimental literature on localization of function[8,10,31] which places functions dependent on egocentric localization such as those are tapped in self-ordering and finger-tapping tasks in more dorsolateral regions while language processing occurs in more ventral locations of the human frontal lobe.

CONCLUSION

Studies in nonhuman primates are beginning to model the basic processes that are central to cognitive operations in primates, including humans. The modular parallel organization of memory circuits in the macaque cerebral cortex suggests a neural basis for similarly modular and segregated cognitive processing systems demonstrable in humans by behavioral analyses, neuropsychological deficits following relatively circumscribed injuries and noninvasive imaging of the brain during performance.

It is important to underscore that while the domains of information processing are modular and parallel, the process carried out within these systems is complex, integrative, and temporally regulated. An outstanding feature of the memory cells of the prefrontal cortex is that a single cell responds differentially at different time points as the process unfolds within a trial, in the terminology of Simon,[30] "within a production." Assuming conservatism in evolution of cortical structure and function, experimental studies in experimental animals may help to decide controversial issues in cognitive psychology and may also shed light on the phylogenetic origins and neural basis of intelligence. Currently, the comparative analysis of working memory functions across monkeys and humans favors the idea of multiple working memory domains, i.e., multiple special purpose systems organized in parallel rather than supporting the concept of a central processor or central executive to account for the diversity and complexity of the human thought process.

REFERENCES

1. Baddeley, A. *Working Memory.* Oxford: London: University Press, 1986.
2. Baddeley, A. D., and G. Hitch. "Working Memory." In *The Psychology of Learning and Motivation*, edited by G. H. Bower. Advances in Research and Theory, Vol. 8, 47–89. New York: Academic Press, 1974.

3. Damasio, A. R. "The Frontal Lobes." In *Clinical Neuropsychology*, edited by K. M. Heilman and E. Valenstein. New York: Oxford University Press, 1979.

4. Fulton, J. F. *Frontal Lobotomy and Affective Behavior.* W. W. Norton, 1951.

5. Funahashi, S., C. J. Bruce, and P. S. Goldman-Rakic. "Mnemonic Coding of Visual Space in the Monkey's Dorsolateral Prefrontal Cortex." *J. Neurophysiol.* **61** (1989): 1–19.

6. Frith, C. D., K. J. Friston, P. F. Liddle, and R. S. J. Frackowiak. "Willed Action and the Prefrontal Cortex in Man: A Study with PET." *Proc. R. Soc. Lond. B.* **244** (1991): 241–246.

7. Funahashi, S., C. J. Bruce, and P. S. Goldman-Rakic. "Dorsolateral Prefrontal Lesions and Oculomotor Delayed-Response Performance: Evidence for Mnemonic Scotomas." *J. Neurosci.* **13(4)** (1993): 1479–1497.

8. Fuster, J. M. *The Prefrontal Cortex*, 2nd ed. New York: Raven Press, 1989.

9. Gaffan, D. "Interaction of Temporal Lobe and Frontal Lobe Memory." This volume.

10. Goldman-Rakic, P. S. "Circuitry of Primate Prefrontal Cortex and Regulation of Behavior by Representational Memory." In *Handbook of Physiology, The Nervous System, Higher Functions of the Brain*, edited by F. Plum, sect. I, vol. V, pt. 1, chap. 9, 373–417. Also in *Am. Physiol. Soc.* (1987): 373–417.

11. Goldman-Rakic, P. S., S. Funahashi, and C. J. Bruce. "Neocortical Memory Circuits." *Qtr. J. Quant. Biol.* **55** (1991): 1025–1038.

12. Guitton, D., H. A. Buchtel, and R. M. Douglas. "Frontal Lobe Lesions in Man Cause Difficulties in Suppressing Reflexive Glances and in Generating Goal-Directed Saccades." *Exp. Brain Res.* **58** (1985): 455–472.

13. Just, M. A., and P. A. Carpenter. "Cognitive Coordinate Systems: Accounts of Mental Rotation and Individual Differences in Spatial Ability." *Psych. Rev.* **92** (1985): 137–172.

14. Kubota, K., and H. Niki. "Prefrontal Cortical Unit Activity and Delayed Alternation Performance in Monkeys." *J. Neurophysiol.* **34** (1971): 337–341.

15. Kubota, K., and S. Funahashi. "Direction-Specific Activities of Dorsolateral Prefrontal and Motor Cortex Pyramidal Track Neurons During Visual Tracking." *J. Neurophysiol.* **47** (1982): 362–376.

16. Mikami, A, S. Ito, and K. Kubota. "Visual Response Properties of Dorsolateral Prefrontal Neurons During a Visual Fixation Task." *J. Neurophysiol.* **47** (1982): 593–605.

17. Mishkin, M. "Perseveration of Central Sets After Frontal Lesions in Monkeys." In *The Frontal Granular Cortex and Behavior*, edited by J. M. Warren and K. Akert. New York: McGraw-Hill, 1964.

18. Mishkin, M., and F. J. Manning. "Non-Spatial Memory After Selective Prefrontal Lesions in Monkeys." *Brain Res.* **143** (1978): 313–323.

19. Newell, A., and H. A. Simon. *Human Problem Solving.* Englewood Cliffs, NJ: Prentice Hall, 1972.

20. O Scalaidhe, S. P., F. A. W. Wilson, and P. S. Goldman-Rakic. "Neurons in the Prefrontal Cortex of the Macaque Selective for Faces." *Soc. Neurosci. Abstr.* **18** (1992): 705.

21. Quintana, J., J. Yajeya, and J. M. Fuster. "Perfrontal Representation of Stimulus Attributes During Delay Tasks. I. Unit Activity in Cross-Temporal Integration of Sensory and Sensory-Motor Information." *Brain Res.* **474** (1988): 211–221.

22. Passingham, R. E. "Delayed Matching After Selective Prefrontal Lesions in Monkeys (*Macac mulatta*)." *Brain Res.* **92** (1975): 89–102.

23. Petrides, M. "Functional Specialization Within the Dorsolateral Frontal Cortex for Serial Order Memory." *Proc. R. Soc. Lond. B* **246** (1991): 293–298.

24. Petrides, M., and B. Milner. "Deficits on Subject-Ordered Tasks After Frontal- and Temporal-Lobe Lesions in Man." *Neuropsychologia* **20** (1982): 249–262.

25. Petrides, M., B. Alivisatos, E. Meyer, and A. C. Evans. "Functional Activation of the Human Frontal Cortex During the Performance of Verbal Working Memory Tasks." *Proc. Natl. Acad. Sci. USA.* **90** (1993): 878–882.

26. Petrides, M., B. Alivisatos, A. C. Evans, and E. Meyer. "Dissociation of Human Mid-Dorsolateral from Posterior Dorsolateral Frontal Cortex in Memory Processing." *Proc. Natl. Acad. Sci. USA.* **90** (1993): 873–877.

27. Rajkowska, G. and P. S. Goldman-Rakic. "Prefrontal Areas 9 and 46 in the Human Cortex: I. Cytoarchitectonic Remapping Using Quantitative Criteria." *Cereb. Cortex* in press.

28. Rosenkilde, C. E., R. H. Bauer, and J. M. Fuster. "Single Cell Activity in Ventral Prefrontal Cortex of Behaving Monkeys." *Brain Res.* **209** (1981): 275–294.

29. Shallice, T. *From Neuropsychology to Mental Structure.* New York: Cambridge University Press, 1988.

30. Simon, H. A. "Near Decomposability and Complexity: How a Mind Resides in a Brain." This volume.

31. Stuss, D. T., and D. F. Benson. *The Frontal Lobes.* New York: Raven Press, 1986.

32. Suzuki, H., and M. Azuma. "Topographic Studies on Visual Neurons in the Dorsolateral Prefrontal Cortex of the Monkey." *Exp. Brain Res.* **53** (1983): 47–58.

33. Wilson, F. A. W., S. P. O Scalaidhe, and P. S. Goldman-Rakic. "Areal and Cellular Segregation of Spatial and of Feature Processing by Prefrontal Neurons." *Soc. Neurosci. Abstr.* **18** (1992): 705.

34. Wilson, F. A. W., S. P. O Scalaidhe, and P. S. Goldman-Rakic. "Dissociation of Object and Spatial Processing Domains in Primate Prefrontal Cortex." *Science* **260** (1993): 1955–1958.

Larry R. Squire and Barbara J. Knowlton
University of California at San Diego, Department of Psychiatry, La Jolla, CA 92093

The Organization of Memory

This chapter is an expanded version of one that appeared originally in *The Cognitive Neurosciences*, edited by M. Gazzaniga, (Cambridge, MA: MIT Press, 1994).

MULTIPLE FORMS OF MEMORY

A major theme in current studies of both humans and experimental animals is that memory is not a single entity but is composed of several separate systems.[109,127,144] Although this idea has a long history,[7,73,105,123,134] experimental work in favor of this view accumulated especially in the 1980s.[17,47,69,90,135,143] One major distinction is between declarative memory, which depends on the integrity of the hippocampus and related structures, and nondeclarative memory, which depends on other brain systems. This distinction between forms of memory is based on neuropsychological studies of amnesic patients, on dissociations in normal subjects showing that conscious memory for recently acquired facts and events is distinct from other

kinds of memory (e.g., skills, habits, priming, and simple conditioning), and dissociations in experimental animals involving lesions or drugs. The terms "explicit"
and "implicit" memory also capture this distinction, but these terms have been
used primarily in the context of human memory.[105] The term "procedural" has
also been used to describe some kinds of nondeclarative memory, particularly skill
learning. The term "nondeclarative" is broader than "procedural," however, and is
meant to include all forms of learning and memory that are not declarative such
as priming, conditioning, and habituation. Figure 1 illustrates a way of classifying
kinds of memory.

Declarative memory is distinct from nondeclarative memory with respect to the
kind of information processing that is involved.[18,113] Declarative memory refers to
memory for facts and events. It is well suited to storing arbitrary associations after
a single trial. Nondeclarative memories are generally acquired gradually across multiple trials. (There are exceptions such as priming and taste aversion conditioning,
which can occur after a single trial). Declarative knowledge is also flexible and can
be readily applied to novel contexts. Nondeclarative memory tends to be inflexible,
bound to the learning situation, and not readily accessed by response systems that
did not participate in the original learning.

The most compelling evidence for this property of declarative and nondeclarative memory systems has come from studies of experimental animals. For example,
normal rats exhibit positive transfer when the elements of previously learned odor
discrimination pairs are recombined to form new pairs. That is, when the reinforced
odor from one pair is presented along with the unreinforced odor from a second pair,

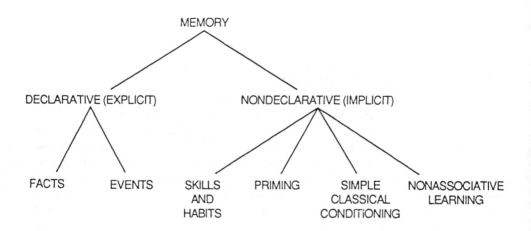

FIGURE 1   A taxonomy of long-term memory. Modified from L. R. Squire, and S. Zola-
Morgan.[126]

normal rats choose the previously reinforced odor and maintain performance at a high level. Rats with lesions to the fimbria-fornix do not exhibit this transfer and perform as if the recombined odor pair is a new problem.[23] Similar findings have been obtained in a study of monkeys with bilateral damage to the hippocampus or related structures.[102]

There are some indications that declarative memory in humans is also more flexible than nondeclarative memory. In one study, amnesic patients were gradually able to learn and use several computer commands, and the patients eventually reached the same performance level as normal subjects. The patients were then impaired relative to normal subjects in applying their knowledge to new situations.[29] In another study, amnesic patients learned to complete simple sentences and then were just as able as control subjects to recall the material when the cues were changed.[115] In this case, the amnesic patients may have depended on residual declarative memory to perform the task. The issue of what characteristics distinguish declarative and nondeclarative memory is an important topic for further study. In the meantime, the link between impaired declarative memory and amnesia is strong enough that amnesic patients provide one method for determining what kind of memory is involved in a behavioral task.[74]

A more recent study of normal subjects found a difference between recognition memory and priming (declarative and nondeclarative memory) that appears, at first glance, to contradict the notion that declarative memory is more flexible than nondeclarative memory.[19] When subjects were asked to recognize recently presented drawings of novel three-dimensional objects (a declarative memory test), they were less accurate at recognizing drawings that had undergone a size change or a mirror reflection transformation as compared to drawings displayed exactly as they were first presented. However, when subjects were asked to judge whether briefly displayed drawings represented possible or impossible three-dimensional objects (a nondeclarative memory test), the facilitation in accuracy produced by prior study of the drawings was not affected by changes in size or mirror reflection. These results might be taken to mean that recognition memory is less flexible than priming. Yet, by a different view, recognition memory is more flexible than priming because all distinctive properties of the objects are represented and are potentially available for retrieval, whereas perceptual priming is limited by an encoding process concerned with structural descriptions of objects that are invariant over size and reflection. Thus, priming is tied to the properties and functions of the cortical regions that process the stimulus, while the representation formed by the hippocampus and related structures provides a full description of the stimulus, including contextual information about the training episode.

THE DISTINCTION BETWEEN SHORT-TERM AND LONG-TERM MEMORY

Information in declarative memory is initially independent of the hippocampus and related structures. In human amnesia, digit span and other measures of short-term (immediate) memory are fully intact.[3,12] This dissociation between short-term and long-term memory can be observed in tests of both verbal and nonverbal materials, including tests of spatial short-term memory.[12] In contrast, cortical lesions that interfere with a particular kind of processing, such as phonological or visuospatial processing, result in a short-term memory deficit specific to that kind of information as well as a corresponding deficit in long-term memory for the same kind of information.[4,5] Thus, declarative memory appears to be organized serially, and information in short-term memory is transformed into persistent, long-term memory by a process that depends on the integrity of the hippocampus and related structures.

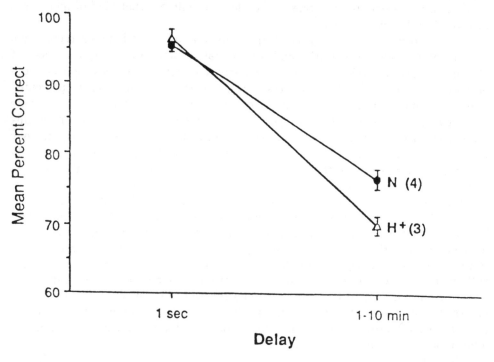

FIGURE 2   Percent of correct responses on the delayed nonmatching to sample task for four normal monkeys (N) and three monkeys with damage to the hippocampal formation (H^+). Performance of the two groups was identical at the 1-sec delay, but differed at longer delays.

The distinction between short-term and long-term memory is also a feature of memory in experimental animals.[58,147] For example, lesions of the dorsal hippocampus in rats spared the recency portion of the serial position curve, while impairing the primacy portion.[58] Moreover, monkeys with bilateral medial temporal lobe lesions acquired the delayed nonmatching to sample task normally when the delay between the presentation of the sample and the choice test was short (1 second or less).[2,89] When trials involving delays from 0.5 sec to 10 min were presented in a mixed fashion, the monkeys performed normally at the shortest delay but were impaired at the longer delays[2] (see Figure 2). These findings together demonstrate that the behavioral impairment in experimental animals following damage to the hippocampus and related structures is a memory problem, not an impairment in perception, rule learning, or some other cognitive function.[98] Indeed, all available evidence supports the conclusion that rats, monkeys, and other animals with damage to the hippocampus and related structures provide a good animal model of human amnesia.[127,150]

RETROGRADE AMNESIA

The brain system that supports declarative memory has only a temporary role in the formation of long-term memory. The pertinent data for this conclusion come from studies of retrograde amnesia, which refers to the loss of memories that were acquired prior to the onset of amnesia. In human amnesia, it has been known for a long time that retrograde amnesia is usually temporally graded, such that recent memories are lost more easily than remote memories.[97] Retrograde amnesia can sometimes be ungraded and extensive, as in conditions such as encephalitis and head trauma, when damage typically occurs beyond the brain system that supports declarative memory.[20,138] Nevertheless, in patients with restricted damage within the hippocampal formation, such as patient R.B., retrograde amnesia is brief, perhaps covering a year or two, at the most prior to the onset of amnesia.[148] Other patients, who presumably have more extensive damage within the medial temporal lobe, have temporally limited retrograde amnesia that extends back one to two decades.[124]

Determining the extent of retrograde amnesia in humans is difficult because the tests are necessarily retrospective, and it is difficult to sample past time periods in an equivalent way. The possibility of studying retrograde amnesia in experimental animals has clarified matters considerably, because the time of acquisition and the strength of memories can be controlled prospectively. In one study, monkeys learned 100 discrimination problems, 20 each at approximately 16, 12, 8, 4, and 2 weeks before removal of the hippocampal formation bilaterally. In one-trial performance tests administered after surgery, the monkeys with lesions were impaired on the discriminations that were learned 2 or 4 weeks before surgery. By contrast, they remembered the discriminations learned longer before surgery as well as normal monkeys. Moreover, the monkeys with lesions remembered the discriminations learned long before surgery significantly better than the discriminations learned just prior to surgery[150] (see Figure 3). These results indicate that early memories are preserved following hippocampal damage because they become independent of the hippocampal formation as time passes after learning.

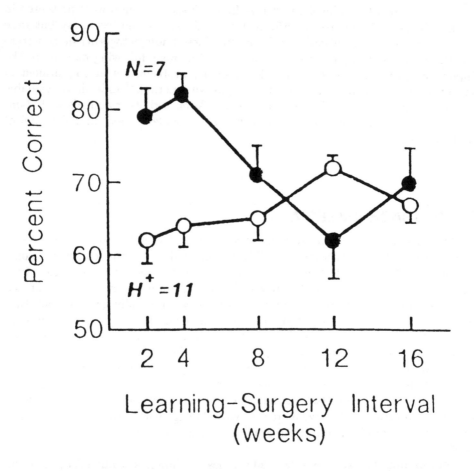

FIGURE 3 The effect of hippocampal formation lesions on retrograde amnesia. Monkeys with lesions were impaired in remembering information learned 2 to 4 weeks before surgery, but remembered objects learned long ago as well as normal monkeys. $N = 7$ normal monkeys, $H^+ = 11$ monkeys with damage to the hippocampal formation. Brackets show standard error of the mean. Modified from S. Zola-Morgan and L. R. Squire.[150]

Similar results using different tasks have also been obtained in rats and mice.[15,59,146] Together, these results demonstrate that retrograde amnesia is temporally graded. One could suppose that temporally graded amnesia occurs because hippocampal damage selectively affects a class of memories that are used for only a short time and spares a class of memories that are destined to endure. If this were the case, however, animals with hippocampal damage should not remember the remote past better than the recent past (for additional discussion, see Zola-Morgan and Squire[150]). Accordingly, the results provide evidence instead for a gradual process of organization and consolidation whereby memory eventually becomes independent of the medial temporal lobe. The medial temporal lobe is not the permanent site of memory storage, but it is necessary for a period of time to allow memories to become stable elsewhere, presumably in neocortex. The medial temporal lobe is the target of highly processed information originating from a variety of cortical regions, and it returns projections to these same cortical regions. The hippocampal formation may serve to bind together disparate aspects of a memory and distill them into a coherent memory trace that subsequently can be accessed by many routes. One possibility is that the medial temporal lobe is the exclusive site of long-term memory storage until a cortical representation is fully developed. Alternatively, the medial temporal lobe may store conjunctions that tie distributed memory storage sites together until more permanent cortico-cortical connections are formed. Computational models of hippocampal-cortical interactions and single-unit studies of the dynamic properties of cortical long-term memory representations will be needed to decide between these alternatives.

RECALL AND RECOGNITION MEMORY

Amnesic patients perform poorly on conventional memory tasks that assess recall or recognition. There has been uncertainty as to whether recall and recognition depend similarly on declarative memory and on the brain system damaged in amnesia or whether recognition performance (but not recall) is based to a significant degree on nondeclarative memory. For example, it is known that recently presented items are processed more quickly and more accurately when the same items are presented a second time. By one view, this improved fluency (e.g., the phenomenon of priming) might lead to a sense of familiarity and thereby influence recognition memory judgments.[27,46,48,70]

If recognition memory performance is derived substantially from nondeclarative processes like perceptual fluency, which are intact in amnesia, then amnesic patients should perform better on recognition tests than would be predicted from their performance on recall tests. A disproportionate advantage for recognition in comparison to recall has been reported previously for memory-impaired patients in two studies that assessed recall and recognition at a single comparison point.[44,45]

However, several of these patients had become amnesic as the result of a ruptured anterior communicating artery aneurysm, a condition known to result in frontal lobe dysfunction. Frontal lobe pathology can affect recall ability more than recognition, presumably because recall requires more effortful search of the contents of memory.[52]

A more recent study of amnesic patients, involving patients with confirmed damage to the medial temporal lobe or diencephalon, found that recall and recognition were affected similarly.[36] In this study, both recall and recognition performance were evaluated across a range of retention intervals so that performance could be compared at several different points. The available results suggest that recall and recognition are equally dependent on declarative memory and that recall is additionally distinguished from recognition by its dependence on the frontal lobes. The view that recognition performance derives little, if any, benefit from nondeclarative memory is also supported by findings that amnesic patients can sometimes perform at chance levels on measures of recognition memory, while at the same time priming is fully intact.[12,122]

Although nondeclarative memory may not affect recognition memory judgments in typical recognition tests, it is possible that item fluency, that is, the process that supports priming, could influence recognition judgments under some circumstances. For example, if stimuli were shown very briefly and if declarative memory for the items were poor, perceptual fluency might influence the decision whether an item had been seen before.[54,55]

THE DISTINCTION BETWEEN REMEMBERING AND KNOWING

A related distinction that is sometimes drawn within recognition memory contrasts the experiences of remembering and knowing. When a item evokes a conscious recollection involving the learning situation, and the recollection includes specific information about the item, a subject is said to "remember" (R). When a subject has only a general feeling of familiarity about a previously presented item, without consciously recollecting anything specific about when the item was first seen, the subject is said to experience "knowing" (K).[137] The subject may be confident an item was seen before, but be unable to remember anything at all about the item in the learning situation.

In some respects, the distinction between remembering and knowing is similar to the declarative/nondeclarative distinction, and R and K responses can be dissociated in a number of ways that are reminiscent of that distinction. For example, R responses are sensitive to level-of-processing effects, while K responses are not.[27] In addition, the frequency of R responses are reduced when items are acquired during divided attention, while K responses are not affected.[28] Both of these manipulations typically affect declarative memory (e.g., recognition memory) more than they affect priming.

It is also possible that both remembering and knowing are dependent on the limbic/diencephalic brain structures that support declarative memory, but that remembering depends additionally on other brain systems. Specifically, R responses may depend on source memory, i.e., memory for when and where a remembered item was learned, whereas K responses are independent of source memory. Source memory depends, at least in part, on the integrity of the frontal lobes.[51,103] Thus, one possibility is that R and K responses are both products of declarative memory and dependent on limbic/diencephalic structures, and that R responses depend additionally the frontal lobes. In one study, elderly subjects with age-appropriate memory abilities generated fewer R responses and more K responses than young adults.[91] The frequency of R responses in the elderly subjects correlated negatively with signs of frontal lobe dysfunction.

The idea that both R and K responses depend on declarative memory is also supported by two other recent findings. First, event-related potentials (ERPs) from old items that elicited R responses were similar to ERPs from old items that elicited K responses until 500 msec after each item was presented.[121] Yet, items that were endorsed as old (i.e., all items that received R or K responses) could be distinguished from items that were endorsed as new beginning about 350 msec after item presentation. Moreover, electrical activity in the hippocampal formation during recognition memory performance appears to be most closely related to task performance during the period 400–500 msec after item presentation,[42,121] suggesting that both R and K responses result from a common process of recollection, dependent on declarative memory and the hippocampus and related structures. The distinction between R and K response arises from a post-recollective process, when subjects attend to the products of their retrieval efforts.

A second finding supporting the idea that both R and K responses are products of declarative memory is that both R and K responses are significantly reduced in amnesic patients.[65] These data also make the more general point that it is difficult to interpret dissociations in the performance of normal subjects (e.g., dissociations resulting from levels-of-processing effects or divided attention manipulations) in terms of processes or distinctions that derive from brain-systems analyses. Some dissociations in the performance of normal subjects appear to reflect the distinction between declarative and nondeclarative memory (e.g., certain effects of superficial vs. deep levels of processing during encoding) and other dissociations in normal subjects appear to reflect the distinction between that aspect of declarative memory that depends on limbic-diencephalic brain structures and that aspect of declarative memory that depends on the frontal lobes (e.g., certain effects of focused vs. divided attention, differences between recall and recognition, and differences between remembering and knowing).

THE DISTINCTION BETWEEN EPISODIC AND SEMANTIC MEMORY

The proposed contribution of frontal lobe function to source memory, and the remembering-knowing distinction, is also relevant to the well-known distinction between episodic and semantic memory. Episodic memory refers to autobiographical memory for events that have a particular spatial and temporal context. Semantic memory refers to factual memory about the world, including factual memory that derives from particular events.[136] The similarity between the concepts of source memory and episodic memory, and the link between source memory and the frontal lobes, suggest that the frontal lobes are more important for episodic memory than for semantic memory.

In amnesia, both episodic and semantic memory are impaired.[24,88,114,119] An important question is whether episodic memory might be relatively more affected in amnesia than semantic memory. Unfortunately, this issue has been difficult to settle. First, it is difficult to compare episodic and semantic memory, because episodic memory by definition is unique to time and place and cannot be repeated. Accordingly, in amnesic patients, the ability to acquire some semantic memory through repetition will always exceed the ability to acquire episodic memory. Second, depending on how one defines semantic memory, there are domains of semantic memory that are severely affected in amnesia (e.g., the ability to learn new facts), and there are domains of semantic memory that are relatively preserved (e.g., the capacity for the gradual learning of artificial grammars and other abilities; see section on nondeclarative memory).

In one study, it was suggested that semantic memory is relatively preserved in amnesia. The amnesic patient K.C., who reportedly has virtually no episodic memory, was able to learn simple sentences across multiple test sessions, using a procedure that minimized interference from the incorrect responses that had been produced on previous trials.[39,139] One way to understand this apparent dissociation between episodic and semantic memory is that patient K.C. became amnesic following head trauma, a condition commonly associated with damage to both the frontal lobe and temporal lobe. Interestingly, a more recent study found that the severely amnesic patient H.M., who has surgical damage to the medial temporal lobe, did not exhibit successful semantic learning, although the testing procedure used for H.M. was similar to the one used for K.C.[140] These results suggest that frontal lobe damage can impair episodic memory more than semantic memory. At the same time, it should be emphasized that it is methodologically difficult to demonstrate a dissociation of episodic and semantic memory in memory-impaired patients (see Squire, Knowlton, and Musen[130] for additional discussion).

THE FRONTAL LOBES, THE DIENCEPHALON, AND THE MEDIAL TEMPORAL LOBE

Patients with lesions involving the frontal lobes have a variety of deficits that affect performance, such as impaired source memory,[103,114] impaired metamemory, i.e., impaired ability to make judgments and predictions about one's own memory ability,[51] impaired memory for temporal order,[51,76] and impaired recall abilities.[52] Diencephalic amnesic patients with Korsakoff's syndrome typically exhibit frontal lobe damage in addition to medial diencephalic damage. The presentation of amnesia in Korsakoff's syndrome is therefore somewhat different than in amnesia resulting from other etiologies.[51,118]

Other than the cognitive deficits attributable to frontal lobe damage, there is striking similarity between diencephalic amnesia and medial temporal lobe amnesia. Both groups have similar forgetting rates within long-term memory[74] and similar spatial memory abilities.[11] The similarity between diencephalic and medial temporal lobe amnesia presumably reflects the close anatomical connections between the diencephalic midline and the medial temporal lobe, and suggests that these two regions can be considered to belong to a single functional system. The two regions undoubtedly make somewhat different contributions to memory, but from the perspective of behavioral criteria, the similarities are more prominent than the differences.

NONDECLARATIVE MEMORY

Despite the severe impairment in recall and recognition exhibited by amnesic patients, a number of memory abilities are intact. These abilities are independent of the limbic/diencephalic structures that support recall and recognition, and depend instead on several different brain systems. The term nondeclarative is used here to refer to this collection of abilities.

PRIMING

Priming refers to the increased ability to identify or detect a stimulus as a result of its recent presentation. The first presentation of an item results in a representation of the stimulus, which can then be subsequently accessed more readily than stimuli that have not been presented previously. In a variety of paradigms, amnesic patients exhibit fully intact priming effects (for recent reviews, see Schacter et al.[108]

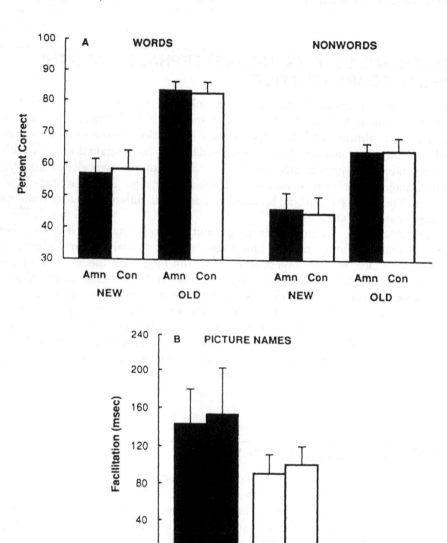

FIGURE 4 Intact priming in amnesic patients on two different tests. (a) Percent of words and nonwords correctly identified in a perceptual identification task. Old items had been presented once previously, and the advantage for identifying old items compared to new items indicates priming. (b) Facilitation of picture naming at two days (2d) and seven days (7d) after a single presentation of the pictures. The facilitation score was obtained by subtracting the time required to name 50 old pictures from the time required to name 50 new pictures. Brackets show standard errors of the mean. AMN, amnesic patients; CON, control subjects. Modified from C. Cave, and L. R. Squire,[12] and F. Haist, G. Musen, and L. R. Squire.[35]

and Squire et al.,[130] Figure 4). This result has been obtained for word and picture priming as measured by word-stem completion,[31] perceptual identification,[13] lexical decision,[120] object naming,[12] and speeded reading.[81] It is also important to note that intact priming in amnesic patients has been demonstrated for novel materials that have no preexisting representations, such as novel objects,[106] nonwords,[35,82] and line patterns.[25,83] These results indicate that priming is not derived simply by activating stored memory representations, but rather is based on the sensory-perceptual traces created by stimulus presentation. The perceptual nature of priming is demonstrated by the fact that the congruence of study and test items is extremely important for priming, and that changes in sensory modality, changes in the voice of the speaker, and even changes in the visual display of the material, can diminish priming.[12,32,46,99,107]

Presentation of items can also influence preferences and judgments about the items. For example, melodies that had been presented before were preferred over new melodies.[53] In addition, both amnesic patients and normal subjects were more likely to judge a proper name as famous if the name had been presented previously.[49,50,86,128] In one study, only nonfamous names were presented first, then subjects were informed that all the names were nonfamous. Subjects were then asked to judge new famous names together with old and new nonfamous names. Amnesic patients continued to exhibit a fame-judgment bias in this circumstance, but normal subjects did not.[131] Normal subjects were able to suppress this effect because they could draw on declarative memory to recall that the items that were just presented were infamous.

The anatomical locus of priming is probably the neocortex. Studies using positron emission tomography (PET) found a reduction in activation within the right posterior neocortex during word-stem completion priming, consistent with the idea that primed items are processed more fluently than new items.[129] Another study of normal subjects involving presentation of word stems in either the left or right visual field identified a right hemisphere locus for word-stem completion priming.[71] Priming was greater when word stems were presented to the left visual field than when word stems were presented to the right visual field. The finding of a right posterior locus suggests that word-stem completion priming relies importantly on visual, orthographic features of the presented material. Priming across modalities, auditory priming, and priming of semantic information, presumably depend on other cortical regions.

Priming is often divided into two types, perceptual and conceptual priming. Perceptual priming is based on the perceptual features of an item and is often sensitive to changes in the surface features of the stimulus.[9,33,71] Conceptual priming is based on semantic information about the stimulus.[32,37,132] For example, in category-exemplar priming, subjects tend to name previously presented examples when asked to generate items from particular categories. Although both types of priming are intact in amnesia, perceptual and conceptual priming can be dissociated. Thus, patients with Alzheimer's disease were impaired on tasks of conceptual priming, but not on tasks of perceptual priming.[56] Recently, the opposite dissociation was demonstrated in a patient with posterior cortical lesions.[57] These dissociations are probably based on the different cortical regions that are important for the two kinds of priming.

Although priming can be long-lasting and can result in new representations, priming is nevertheless limited in comparison to declarative memory. While declarative memory is well suited for forming new associations between arbitrary stimuli in a single trial, in nondeclarative memory robust associations are formed more gradually. For example, priming of new associations was weak or absent in amnesic patients when word stems were presented at test alongside a word that had been paired with the target word. It appears that declarative memory is necessary for this phenomenon to occur.[104,116] In another study involving elderly subjects, including memory-impaired subjects, word pairs that had been presented previously were read faster than word pairs formed by recombining previously presented pairs.[78] However, in a more recent study this effect was observed only when word pairs were presented multiple times.[85] In three different experiments, one-trial implicit learning of novel associations did not occur for either normal subjects or amnesic patients.

In another study, both amnesic patients and normal subjects exhibited implicit learning of the association between a word and the color in which it was displayed. This effect was demonstrated using the Stroop paradigm, in which color names were printed in incongruent colors (e.g., the word red was printed in the color green). Naming times for colors decreased across multiple trials when the words were presented in consistent incongruent colors, and naming times subsequently increased when the words were presented in new incongruent colors.[84] The associations between color names and the incongruent colors in which they were printed were not available to declarative memory, as indicated by the fact that subjects performed at chance on a recognition memory test that asked about the pairings.

In summary, completely normal priming has been demonstrated in amnesic patients in a number of different tasks. These effects occur despite entirely normal baseline performance as measured by both speed and accuracy.[31,35,38,81,82] At this point the evidence strongly supports the distinction between declarative memory and priming. As more is learned about the anatomical structures that support priming, this distinction will be based on structural as well as functional evidence.

HABITS AND SKILLS

A habit is a tendency or disposition to engage in a behavior in response to a specific set of stimuli. A skill is a procedure (motor, perceptual, or cognitive) for operating in the world. Skill learning often draws on declarative memory as well as nondeclarative memory, especially in complex learning situations where subjects can benefit from conscious mediation. Nevertheless, in some circumstances skill learning is largely nondeclarative as evidenced by the fact that amnesic patients learn at an entirely normal rate. This has been demonstrated for rotor-pursuit learning,[10] reading mirror-reversed text,[17] and adaptation-level effects based on lifting and judging weights.[6]

Amnesic patients can also learn normally when the information to be acquired is not exclusively perceptual or motor. In one study, subjects performed a serial reaction-time task in which they responded successively to a sequence of four illuminated spatial locations.[87] The task was to press one of four keys as rapidly as possible, as soon as the location above that key was illuminated. Amnesic patients and normal subjects exhibited learning of a repeating sequence as indicated by decreasing reaction time for key presses as the sequence repeated itself. When the sequence was changed, reaction times increased again. Subjects were able to learn the sequence, even when they were judged to have no declarative knowledge of it. Subjects were judged to have no declarative knowledge when they were unable to generate the sequence in subsequent tests and were unaware that a sequence had been presented.[87,145] Implicit learning of these sequences was disrupted under divided-attention conditions, although more simple sequences could be learned under divided attention.[16,87] Thus, it appears that automaticity is not a universal feature of implicit learning.

A recent study questioned whether declarative knowledge of the sequence was tested in a sufficiently sensitive way to support the claim that sequence learning had occurred without awareness. In the original studies,[87,145] subjects were either asked to report the spatial sequence verbally or to try to predict each successive element of the sequence by pressing the key in the location where the light would next appear. After each response, the correct location in the sequence was illuminated. In the newer study,[93] declarative knowledge for the sequence was tested by asking subjects to press the keys in the same sequence that had been used during training (free generation task). No feedback was given for responses. With this procedure, subjects were clear about the link between the reaction time task and the generation task, and the lack of feedback during the generate task made it rather similar to the reaction-time task. Also, subjects were asked to discriminate permissible sequences from nonpermissible sequences. Subjects performed above chance on these measures early during training and as early in training as sequence learning could be detected by the reaction-time measure.

One difficulty with measures of declarative knowledge that closely reproduce the conditions of implicit learning tests is that subjects may be able to respond on the basis of nondeclarative knowledge. That is, subjects may be able to monitor their behavior and report on what they do, just as one could observe the sequence of actions that comprise a learned motor skill and describe the sequence with some accuracy. Yet, one's declarative knowledge in this circumstance would be epiphenomenal in that it is not necessary to have such knowledge to acquire the skill. For this reason, the findings from amnesic patients are particularly useful because they provide a tool for assessing the contributions of declarative memory to task performance. A finding that amnesic patients learn and remember entirely normally provides strong evidence that long-term declarative memory is not needed for performance. It is possible that some declarative knowledge develops during initial learning, and that even in amnesic patients such knowledge is supported by their intact immediate memory capacity, but it is a different matter whether declarative knowledge for what is learned in a task can or does persist within long-term memory once learning is completed.

PRODUCTION SYSTEMS

In some cases, skills can involve more abstract information, and what is learned is neither perceptual nor motor. In one task, subjects imagine that they operate a sugar production factory. Subjects select how many workers should be hired in order to achieve an optimal level of sugar production on each trial. The total number of workers hired on the previous trial, the number of workers hired on the current trial, and the output variable (the amount of sugar produced) are related by a simple formula. Subjects gradually learn to respond in order to obtain a desired level of sugar production, although they need not acquire much reportable knowledge about the rule.[8] When more sensitive measures of knowledge are taken, it does appear that subjects acquire some conscious knowledge of the task structure as they improve their performance.[101] Again, it is difficult to determine from such data whether conscious knowledge of the task structure is necessary for performance to improve. Thus, it is important to note that amnesic patients perform as well as normal subjects during the early learning of the sugar-production task[125] With extended training, normal subjects are able to outperform amnesic patients, and normal subjects are also better than amnesic patients at answering questions about task strategy.

PROBABILISTIC CLASSIFICATION LEARNING

Early nondeclarative learning followed by later declarative learning is also observed in skill-learning tasks in which subjects try to predict one of two outcomes based on a set of cues that are probabilistically associated with each outcome. In one such task, subjects try to decide which of two fictitious diseases an imaginary patient has based on a list of one to four symptoms. Each symptom independently predicts one of the two disease outcomes with a particular probability.[30] This task shares formal aspects with classical conditioning. That is, the separate cues compete for associative strength with the outcome in much the same way that conditioned stimuli compete for associative strength with the unconditioned stimulus.[14,30,112] In three different tasks of probabilistic classification learning, amnesic patients improved their classification performance at the same rate as normal subjects.[63] Probabilistic associations may be learned implicitly because information about a single trial is not as useful for performance as information about the probabilistic relationship between cues and outcomes, which necessarily accrues over many trials. Learning occurred at a normal rate for the amnesic patients during approximately the first 50 trials of training, during which subjects improved their classification performance from 50% correct (chance) to about 70% correct. However, with extended training, normal subjects surpassed the performance of amnesic patients, presumably because they were able to decipher the task to some extent and to remember some of the relationships explicitly.

ARTIFICIAL GRAMMAR LEARNING

Another task in which classification judgments about category membership can be based on nondeclarative knowledgeis artificial grammar learning. In artificial grammar learning, subjects first see a series of letter strings, one at a time, which are formed according to a finite-state rule system (see Figure 5). The letter strings are generated by traversing the diagram from the input arrow to an output arrow, adding a letter at each transition. Subjects are told about the underlying rule system only after viewing the letter strings. They are then asked to judge whether new letter strings adhere or do not adhere to the rules. Although subjects are not able to describe the basis for their judgments very well, they are able to classify new letter strings at a level significantly above chance (see Reber,[96] for a review). Because of the disparity between the ability of subjects to classify letter strings correctly and their poor conscious knowledge of the rule system, artificial grammar learning has often been described as implicit. However, it has also been argued that subjects may be using partially valid declarative knowledge of the grammar to make their judgments, and it has been argued that declarative knowledge about the grammar can be elicited from subjects using more sensitive test measures.[22,92] Recent studies have helped to resolve this debate by showing that amnesic patients are able to make classification judgments as well as normal subjects in an artificial grammar learning task, despite their severe impairment in the ability to recognize the particular letter strings that were used to teach the grammar.[62,64]

Separate from the question of whether artificial grammar learning is implicit or not is the question of whether subjects are learning complex, abstract information about the grammatical rule system, or whether they are learning more concrete information about the particular training examples that are shown. For example,

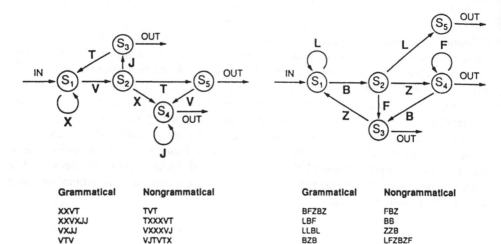

FIGURE 5 Two finite-state rule systems used to generate the letter strings of artificial grammars. Examples of grammatical and nongrammatical letter strings are listed below each rule system. Modified from M. Abrams and A. S. Reber,[1] and B. Knowlton, S. J. Ramus, and L. R. Squire.[62]

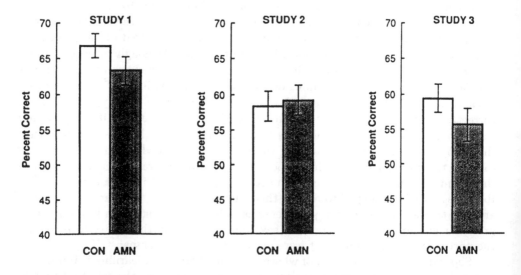

FIGURE 6 The results of three separate studies showing normal performance of amnesic patients (AMN) compared to control subjects (CON) on classification tasks based on artificial grammars. Brackets show standard error of the mean. Modified from B. J. Knowlton, S. J. Ramus, and L. R. Squire,[62] and B. J. Knowlton, and L. R. Squire.[64]

judgments could be based on the overall similarity of individual test items to specific training items,[141] or on the occurrence of features in the test items that are also repeated in the training items.[111] Features such as sequences of two or three letters (chunks) could gain associative strength with the grammatical category as they are repeated across the set of training items. By this view, subjects subsequently endorse test items that contain letter chunks with high associative strength. Recent studies indicate that overall similarity of test items to specific training items does not influence judgments of grammaticality, when chunk strength is equated between similar and nonsimilar test items.[63] Grammaticality judgments presumably depend on abstract knowledge of rules, or on rather concrete information about permissible letter chunks that is accumulated across the training items or on both kinds of knowledge.

PROTOTYPE ABSTRACTION AND CATEGORY LEARNING

In rtificial grammar learning, category membership is defined by whether or not test items adhere to a finite rule system. By contrast, membership in natural categories is usually defined by the extent to which test items resemble a prototype of the category.[100] Thus, subjects will classify the prototype as being a member of the category more readily than other examples even when the prototype itself has never been seen. One possibility is that subjects abstract the prototype from a set of training examples and use this information to make category judgments.[95] Alternatively, category judgments may be based on a comparison of test items to examples stored in declarative memory. By this view, the prototype is classified as belonging to the category because the prototype is usually similar to a large number of examples, just as recognition memory judgments are based on the similarity between test items and training items.[43,75] Studies of amnesic patients should make it possible to distinguish between these possibilities. In a classification task using dot patterns similar to those used by Posner and Keele,[95] amnesic patients exhibited normal classification learning despite a severe impairment at recognizing the specific dot patterns that were presented during training.[61] Amnesic patients and normal subjects were shown distortions of a prototypic dot pattern during training. Both groups subsequently demonstrated that they had abstracted the prototype from the examples. That is, both groups judged the prototype and low distortions of the prototype as members of the training category more often than they judged higher distortions of the prototype and random dot patterns to be members of the training category (see Figures 7 and 8). These results suggest that category-level information is acquired independently of declarative memory for training exemplars. Category-level information could be constructed nondeclaratively (implicitly) either by forming an abstracted prototype or by making comparisons with instances that are stored in implicit memory. Category-level judgments are independent of the ability to remember declaratively the particular instances used during training.

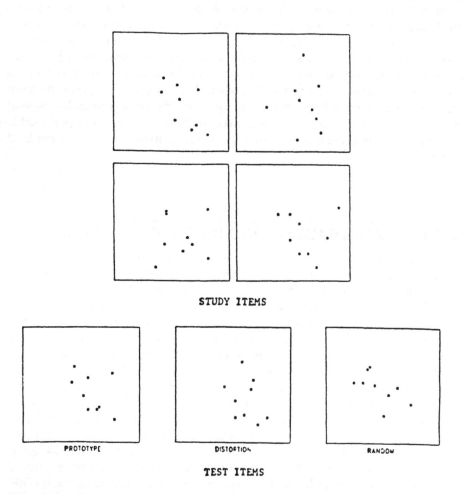

STUDY ITEMS

TEST ITEMS

FIGURE 7 Examples of four study items and three test items used to assess classification learning. The study items are all arithmetic distortions of a prototype dot pattern that subjects do not see. The test items include the prototype pattern, novel distortions of the prototype, and random dot patterns that provide a measure of baseline classification performance.

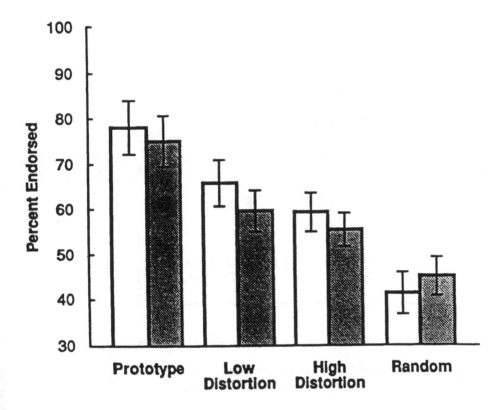

FIGURE 8 Performance on the dot pattern classification task according to type of test item. Open bars = control subjects; shaded bars = amnesic patients. Error bars show standard error of the mean.

FROM MEMORY SYSTEMS TO BRAIN SYSTEMS

Memory is a set of functions, each of which depends on different brain systems. Declarative memory is the product of a unique system that is dependent on limbic/diencephalic structures, which operate in concert with neocortex (see Figure 9). Studies of nonhuman primates have elucidated the brain structures and connections that support declarative memory.[77,126] The important structures in the medial temporal lobe are the hippocampus, the entorhinal cortex, the parahippocampal cortex, and the perirhinal cortex. The amygdala is not part of the medial temporal lobe system for declarative memory.[149]

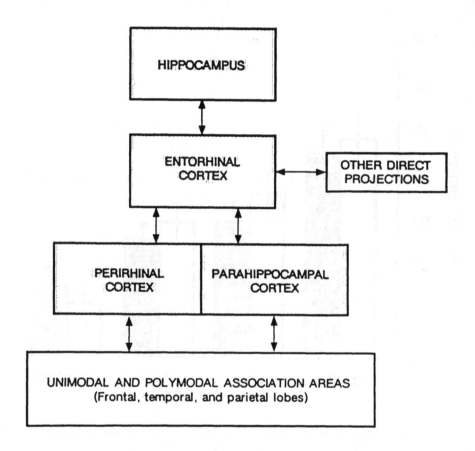

FIGURE 9 A schematic view of the structures and connections important for declarative memory. Shaded areas indicate structures within the medial temporal lobe.

One difficulty with evaluating the contributions to memory of the separate components of the medial temporal lobe is that individual studies usually employ small numbers of animals. Recently, it has become feasible to combine data from several different experiments.[153] Damage to the hippocampal region, caused either by ischemia or radiofrequency lesions, resulted in a significant memory impairment. This level of impairment was increased when the area of the damage was systematically enlarged to include the parahippocampal cortex and posterior entorhinal cortex (the H^+ lesion). In addition, the impairment associated with an H^+ lesion was increased still further when the H^+ lesion was extended forward to include anterior entorhinal cortex and perirhinal cortex (Figure 10).

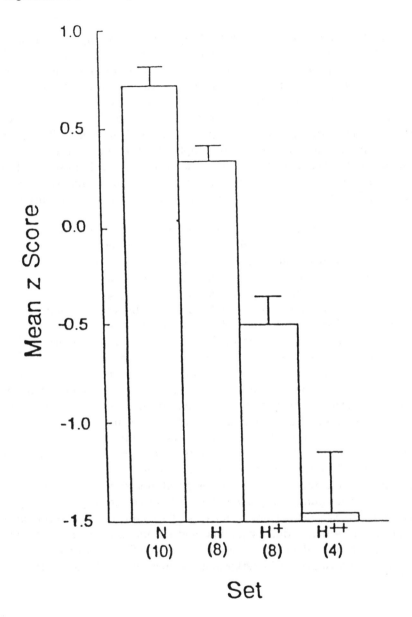

FIGURE 10 Mean z scores based on 4 measures of memory for 10 normal monkeys (N), 8 monkeys with damage to the hippocampal region (H), 8 monkeys with damage that also included the adjacent entorhinal and parahippocampal cortices (H^+), and 4 monkeys in which the $H+$ lesion was extended forward to include the anterior entorhinal cortex and the perirhinal cortex (H^{++}). As more components of the medial temporal lobe memory system were included in the lesion, the severity of memory impairment increased. Brackets show standard errors of the mean.[153]

These findings from monkeys are consistent with the findings from human amnesia. Patient R.B., who sustained a discrete lesion of hippocampal area CA1 following an ischemic episode,[148] had a moderately severe and clinically significant memory impairment. However, R.B.'s impairment was not nearly so severe as the impairment exhibited by patient H.M.[110] The important finding was that the severity of memory impairment associated with medial temporal lobe damage becomes more severe as more of the medial temporal lobe is damaged.[152] The parahippocampal and perirhinal cortices are not simply conduits for sending information to the hippocampus. Damage to the hippocampal region itself causes a relatively mild level of impairment. The fact that memory impairment increases when the adjacent cortical regions are damaged indicates that these areas themselves also contribute to memory function.

One recent study attributed the memory deficit after large medial temporal lobe lesions to damage to the blood supply for the adjacent cortical area TE.[26] Area TE is involved in higher-level visual processing, so that damage to this region would be expected to produce a performance deficit on visual memory tasks such as the delayed-nonmatching-to-sample test. However, several lines of evidence indicate that the impairment after medial temporal lobe lesions is not due to TE damage. First, monkeys with large medial temporal lobe lesions can learn habit-like nondeclarative memory tasks normally, such as the 24-hr concurrent discrimination task. Yet, lesions of area TE impair performance on this task.[94] Second, medial temporal lobe lesions spare visual short-term memory.[2] They do not cause a general deficit in visual performance. Finally, medial temporal lobe lesions impair performance on tactual memory tests and visual memory tests.[79,80] TE lesions, by contrast, produce an impairment only in visual tasks, not in tactual tasks.[79] Thus, damage to the medial temporal lobe memory system, unlike TE damage, results in a multimodal deficit similar to human amnesia. Damage to the medial temporal lobe does not reproduce the effects of area TE damage.

The information processed by medial temporal lobe structures is also directed to areas in the diencephalon important for declarative memory. The important areas appear to be the medial dorsal nucleus of thalamus, the anterior nucleus, the region of the internal medullary lamina, and the mammillary nuclei.[34,151] The development of an animal model of alcoholic Korsakoff's syndrome in the rat[68] provides a new opportunity for investigating the anatomy of diencephalic amnesia.

ANATOMICAL SUBSTRATES OF NONDECLARATIVE MEMORY

The medial temporal lobe, and related diencephalic structures, is not the only brain system involved in learning and memory. Other systems are involved in learning other kinds of information. For example, habit learning likely depends on different brain regions depending on the type of habit that is learned. Other Studies with experimental animals suggest that the caudate nucleus is critical for some forms of stimulus-response habit learning. Caudate lesions in rats, but not fornix lesions, impaired the learning of a win-stay habit in a radial maze, while fornix lesions but not caudate lesions impaired the learning of win-shift tasks.[90] A similar dissociation of declarative memory and habit learning has also been obtained in monkeys. Lesions of the tail of the caudate nucleus, which is a target of projections from area TE, impaired the learning of stimulus-response associations in the 24-hr concurrent task, whereby monkeys learn 20 object pairs concurrently while receiving one trial per day on each of the 20 pairs.[142] By contrast, monkeys with large medial temporal lobe lesions were severely impaired on the delayed nonmatching to sample task, but they learned the 24-hr concurrent task at almost a normal rate.[69]

The neostriatum may also be important for the learning of skills and habits in human subjects. Patients with Huntington's disease were impaired on sensorimotor tasks including the reading of mirror-reversed text,[72] rotor-pursuit learning,[40] adaptation level effects,[41] and sequence learning using a reaction-time task.[60] Although declarative memory is not normal in these patients, the same patients who performed more poorly than amnesic patients on sensorimotor tasks performed better than amnesic patients on tests of declarative memory. Patients with Huntington's disease may be impaired on the sensorimotor tasks because they are deficient at forming motor programs. Accordingly, an important question is whether patients with Huntington's disease would be impaired on the learning of habitlike tasks that do not have a motor component, such as artificial grammar learning or probabilistic classification learning. Alternatively, rather than supposing that these tasks are habitlike and depend on the neostriatum, one could suppose that acquiring category-level knowledge is dependent on neocortex. That is, the processing of exemplars in neocortex might gradually lead to a cortical representation of the commonalities between the training items, and the resulting cortical representation could provide a basis for category-level judgments. Further studies should make it possible to choose among these possibilities.

Other kinds of learning and memory can also be distinguished. The amygdala is part of a circuit specialized for emotional learning, including classical conditioning of emotional responses.[21,67] In addition, the cerebellum and associated brainstem and midbrain nuclei are necessary for classical conditioning of discrete responses of the skeletal musculature.[66,133] Thus, it is possible to link particular brain regions and systems to various kinds of memory (see Figure 11). The unique features of declarative memory, such as rapid learning and conscious, flexible access, are instantiated within the medial temporal lobe—diencephalic circuitry and in the divergent connections between these systems and the neocortex where long-term declarative memories are presumably stored. The next challenge for cognitive neuroscience is to relate functional properties of each memory system to the anatomical and physiological characteristics of the corresponding brain systems.

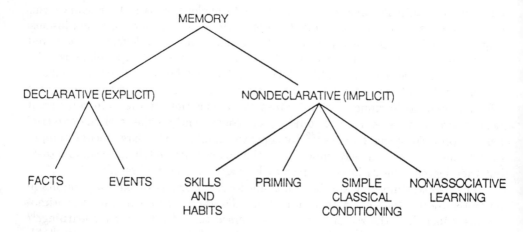

FIGURE 11   A taxonomy of long-term memory and associated brain structures.

REFERENCES

1. Abrams, M., and A. S. Reber. "Implicit Learning in Special Populations." *J. Psycholing. Res.* **17** (1989): 425–439.
2. Alvarez-Royo, P., S. Zola-Morgan, and L. R. Squire. "Impaired Long-Term Memory and Spared Short-Term Memory in Monkeys with Medial Temporal Lobe Lesions: A Response to Ringo." *Behav. Brain Res.* **52** (1992): 1–5.
3. Baddeley, A. P., and E. K. Warrington. "Amnesia and the Distinction Between Long- and Short-Term Memory." *J. Verbal Learn. Verbal Behav.* **9** (1970): 176–189.
4. Baddeley, A. D., and G. J. Hitch. "Working Memory." In *Psychology of Learning and Motivation: Advances in Research and Theory*, edited by G. A. Bower, 47–90. New York: Academic Press, 1974.
5. Baddeley, A., C. Papagno, and G. Vallar. "When Long-Term Learning Depends on Short-Term Storage." *J. Mem. Lang.* **27** (1988): 586–595.
6. Benzing, W. C., and L. R. Squire. "Preserved Learning and Memory in Amnesia: Intact Adaptation-Level Effects and Learning of Stereoscopic Depth." *Behav. Neurosci.* **103** (1989): 538–547.
7. Bergson, H. *Matter and Memory.* London: Allen and Unwin, 1911.
8. Berry, D., and D. Broadbent. "On the Relationship Between Task Performance and Associated Verbalizable Knowledge." *Quart. J. Exp. Psychol.* **36A** (1984): 209–231.
9. Blaxton, T. A. "Investigating Dissociations Among Memory Measures: Support for a Transfer Appropriate Processing Framework." *J. Exp. Psychol. Learn. Mem. Cog.* **15** (1989): 657–668.

10. Brooks, D. N., and A. Baddeley. "What Can Amnesic Patients Learn?" *Neuropsychologia* **14** (1976): 111–122.
11. Cave, C. B., and L. R. Squire. "Equivalent Impairment of Spatial and Nonspatial Memory Following Damage to the Human Hippocampus." *Hippocampus* **1** (1991): 329–340.
12. Cave, C., and L. R. Squire. "Intact and Long-Lasting Repetition Priming in Amnesia." *J. Exp. Psychol. Learn. Mem. Cog.* **18** (1992): 509–520.
13. Cermak, L. S., N. Talbot, K. Chandler, and L. R. Wolbarst. "The Perceptual Priming Phenomenon in Amnesia." *Neuropsychologia* **23** (1985): 615–622.
14. Chapman, G. B., and S. J. Robbins. "Cue Interation in Human Contingency Judgment." *Mem. Cog.* **18** (1990): 537–545.
15. Cho, Y. H., D. Beracochea, and R. Jaffard. "Extended Temporal Gradient for the Retrograde and Anterograde Amnesia Produced by Ibotenate Entorhinal Cortex Lesions in Mice." *J. Neurosci.* **13** (1993): 1759–1766.
16. Cohen, A., R. I. Ivry, and S. W. Keele. "Attention and Structure in Sequence Learning." *J. Exp. Psychol. Learn. Mem. Cog.* **16** (1990): 17–30.
17. Cohen, N. J., and L. R. Squire. "Preserved Learning and Retention of Pattern Analyzing Skill in Amnesia: Dissociation of Knowing How and Knowing That." *Science* **210** (1980): 207–209.
18. Cohen, N. J. "Preserved Learning Capacity in Amnesia: Evidence for Multiple Memory Systems." In *Neuropsychology of Memory*, edited by L. R. Squire and N. Butters. New York: Guilford Press, 1984.
19. Cooper, L. A., D. L. Schacter, S. Ballesteros, and C. Moore. "Priming and Recognition of Transformed Three-Dimensional Objects: Effects of Size and Reflection." *J. Exp. Psychol. Learn. Mem. Cog.* **18** (1992): 43–57.
20. Damasio, A. R., P. J. Eslinger, H. Damasio, G. W. Van Hoesen, and S. Cornell. "Multiple Amnesic Syndrome Following Bilateral Temporal and Basal Forebrain Damage." *Arch. Neurol.* **42** (1985): 252–259.
21. Davis, M. "The Role of the Amygdala in Fear-Potentiated Startle: Implications for Animal Models of Anxiety." *Trends Pharm. Sci.* **13** (1992): 35–41.
22. Dulany, D. E., R. A. Carlson, and G. I. Dewey. "A Case of Syntactical Learning and Judgment: How Conscious and How Abstract?" *J. Exp. Psychol. Gen.* **113** (1984): 541–555.
23. Eichenbaum, H., P. Mathews, and N. J. Cohen. Further Studies of Hippocampal Representation During Odor Discrimination Learning." *Behav. Neurosci.* **103** (1989): 1207–1216.
24. Gabrieli, J. D. E., N. J. Cohen, and S. Corkin. "The Impaired Learning of Semantic Knowledge Following Medial Temporal-Lobe Resection." *Brain Cog.* **7** (1988): 157–177.
25. Gabrieli, J. D. E., W. Milberg, M. M. Keane, and S. Corkin. "Intact Priming of Patterns Despite Impaired Memory." *Neuropsychologia* **28** (1990): 417–427.
26. Gaffan, D., and C. Lim. "Hippocampus and the Blood Supply to TE: Parahippocampal Pial Section Impairs Visual Discrimination Learning in Monkeys." *Exp. Brain Res.* **87** (1991): 227–231.

27. Gardiner, J. M. "Recognition Failures and Free-Recall Failures: Implications for the Relation Between Recall and Recognition." *Mem. Cog.* **16** (1988): 446–451.

28. Gardiner, J. M., and A. J. Parkin. "Attention and Recollective Experience in Recognition Memory." *Mem. Cog.* **18** (1990): 579–583.

29. Glisky, E. L., D. L. Schacter, and E. Tulving. "Learning and Retention of Computer-Related Vocabulary in Memory-Impaired Patients: Method of Vanishing Cues." *J. Clin. Exp. Neuropsy.* **8** (1986): 292–312.

30. Gluck, M. A., and G. H. Bower. "From Conditioning to Category Learning: An Adaptive Network Model." *J. Exp. Psychol. Gen.* **117** (1988): 227–247.

31. Graf, P., L. R. Squire, and G. Mandler. "The Information that Amnesic Patients Do Not Forget." *J. Exp. Psychol. Learn. Mem. Cog.* **10** (1984): 164–178.

32. Graf, P., A. P. Shimamura, and L. R. Squire. "Priming Across Modalities and Priming Across Category Levels: Extending the Domain of Preserved Function in Amnesia." *J. Exp. Psychol. Learn. Mem. Cog.* **11** (1985): 386–396.

33. Graf, P., and L. Ryan. "Transfer-Appropriate Processing for Implicit and Explicit Memory." *J. Exp. Psychol. Learn. Mem. Cog.* **16** (1990): 978–992.

34. Graff-Radford, N. R., D. Tranel, G. W. Van Hoesen, and J. Brandt. "Diencephalic Amnesia." *Brain* **113** (1990): 1–25.

35. Haist, F., G. Musen, and L. R. Squire. "Intact Priming of Words and Nonwords in Amnesia." *Psychobiology* **19** (1991): 275–285.

36. Haist, F., A. P. Shimamura, and L. R. Squire. "On the Relationship Between Recall and Recognition Memory." *J. Exp. Psychol. Learn. Mem. Cog.* **18** (1992): 691–702.

37. Hamann, S. "Level-of-Processing Effects in Conceptually Driven Implicit Tasks." *J. Exp. Psychol. Learn. Mem. Cog.* **16** (1990): 970–977.

38. Hamann, S., and L. R. Squire. "Perceptual Thresholds and Priming in Amnesia." *Neuropsych* (1993): in press.

39. Hayman, C. A. G., C. A. MacDonald, and E. Tulving. "The Role of Repetition and Associative Interference in New Semantic Learning in Amnesia: A Case Experiment." *J. Cognit. Neurosci.* **5** (1993): 375–389.

40. Heindel, W. C., N. Butters, and D. P. Salmon. "Impaired Learning of a Motor Skill in Patients with Huntington's Disease." *Behav. Neurosci.* **102** (1988): 141–147.

41. Heindel, W. C., D. P. Salmon, and N. Butters. "The Biasing of Weight Judgments in Alzheimer's and Huntington's Disease: A Priming or Programming Phenomenon?" *J. Clin. Exp. Neuropsy.* **13** (1991): 189–203.

42. Heit, G., M. E. Smith, and E. Halgren. "Neuronal Activity in the Human Medial Temporal Lobe During Recognition Memory." *Brain* **113** (1990): 1093–1112.

43. Hintzman, D. "Schema Abstraction in a Multiple-Trace Memory Model." *Psychol. Rev.* **93** (1986): 411–428.

44. Hirst, W., M. K. Johnson, E. A. Phelps, G. Risse, and B. T. Volpe. "Recognition and Recall in Amnesics." *J. Exp. Psychol. Learn. Mem. Cog.* **12** (1986): 445–451.
45. Hirst, W., M. K. Johnson, E. A. Phelps, and B. T. Volpe. "More on Recognition and Recall in Amnesics." *J. Exp. Psychol. Learn. Mem. Cog.* **14** (1988): 758–762.
46. Jacoby, L. L. and M. Dallas. "On the Relationship Between Autobiographical Memory and Perceptual Learning." *J. Exp. Psychol. Gen.* **3** (1981): 306–340.
47. Jacoby, L. L., and D. Witherspoon. "Remembering Without Awareness." *Can. J. Psychol.* **32** (1982): 300–324.
48. Jacoby, L. L. "Remembering the Data: Analyzing Interactive Processes in Reading." *J. Verbal Learn. Verbal Behav.* **22** (1983): 485–508.
49. Jacoby, L. L., C. Kelley, J. Brown, and J. Jasechko. "Becoming Famous Overnight: Limits on the Ability to Avoid Unconscious Influences of the Past." *J. Pers. Soc. Psychol.* **56** (1989): 326–338.
50. Jacoby, L. L., V. Woloshyn, and C. Kelley. "Becoming Famous Without Being Recognized: Unconscious Influences of Memory Produced by Dividing Attention." *J. Exp. Psychol. Gen.* **118** (1989): 115–125.
51. Janowsky, J. S., A. P. Shimamura, and L. R. Squire. "Source Memory, Impairment in Patients with Frontal Lobe Lesions." *Neuropsychologia* **27** (1989): 1043–1056.
52. Jetter, W., U. Poser, R. B. Freeman, and J. H. Markowitsch. "A Verbal Long Term Memory Deficit in Frontal Lobe Damaged Patients." *Cortex* **22** (1986): 229–242.
53. Johnson, M. K., J. K. Kim, and G. Risse. "Do Alcoholic Korsakoff's Syndrome Patients Acquire Affective Reactions?" *J. Exp. Psychol. Learn. Mem. Cog.* **11** (1985): 22–36.
54. Johnston, W. A., W. J. Dark, and L. L. Jacoby. "Perceptual Fluency and Recognition Judgments." *J. Exp. Psychol. Learn. Mem. Cog.* **11** (1985): 3–11.
55. Johnston, W. A., K. J. Hawley, and M. G. Elliott. "Contribution of Perceptual Fluency to Recognition Judgments." *J. Exp. Psychol. Learn. Mem. Cog.* **17** (1991): 210–223.
56. Keane, M. M., J. D. E. Gabrieli, A. C. Fennema, J. H. Growdon, and S. Corkin. "Evidence for a Dissociation Between Perceptual and Conceptual Priming in Alzheimer's Disease." *Behav. Neurosci.* **105** (1991): 326–342.
57. Keane, M. M., H. Clarke, and S. Corkin. "Impaired Perceptual Priming and Intact Conceptual Priming in a Patient with Bilateral Posterior Cerebral Lesions." *Soc. Neurosci. Abs.* **18** (1992): 386.
58. Kesner, R. P., and J. M. Novak. "Serial Position Curve in Rats: Role of the Dorsal Hippocampus." *Science* **218** (1982): 173–175.
59. Kim, J. J., and M. S. Fanselow. "Modality-Specific Retrograde Amnesia of Fear." *Science* **256** (1992): 675–677.

60. Knopman, D. S., and M. J. Nissen. "Procedural Learning is Impaired in Huntington's Disease: Evidence from the Serial Reaction Time Task." *Neuropsychologia* **29** (1991): 245–254.

61. Knowlton, B. J., and L. R. Squire. ""The Learning of Natural Categories: Parallel Memory Systems for Item Memory and Category-Level Knowledge." *Science* **262** (1993): 1747–1749.

62. Knowlton, B. J., S. J. Ramus, and L. R. Squire. "Intact Artificial Grammar Learning in Amnesia: Dissociation of Classification Learning and Explicit Memory for Specific Instances." *Psychol. Sci.* **3** (1992): 172–179.

63. Knowlton, B. J., M. A. Gluck, and L. R. Squire. "Probabilistic Category Learning in Amnesia." *Learn. & Mem.* (in press).

64. Knowlton, B. J., and L. R. Squire. "The Information Acquired During Artificial Grammar Learning." *J. Exp. Psychol. Learn. Mem. Cog.* **20** (1994): 79–91.

65. Knowlton, B. J., and L. R. Squire. "Remembering and Knowing: Two Different Expressions of Declarative Memory." Submitted.

66. Lavond, D. G., J. J. Kim, and R. F. Thompson. "Mammalian Brain Substrates of Aversive Classical Conditioning." *Ann. Rev. Psychol.* **44** (1993): 317–342.

67. LeDoux, J. E. "Emotion." In *Handbook of Physiology: The Nervous System V. Higher Functions of the Nervous System*, edited by J. M. Brookhart and V. B. Mountcastle, 419–460. Bethesda, MD: APS, 1987.

68. Mair, R. G., C. D. Anderson, P. J. Langlais, and W. J. McEntree. "Behavioral Impairments, Brain Lesions and Monoaminergic Activity in the Rat Following a Bout of Thiamine Deficiency." *Behav. Brain Res.* **27** (1988): 223–239.

69. Malamut, B. L., R. C. Saunders, and M. Mishkin. "Monkeys with Combined Amygdalo-Hippocampal Lesions Succeed in Object Discrimination Learning Despite 24-Hour Intertrial Intervals." *Behav. Neurosci.* **98** (1984): 759–769.

70. Mandler, G. "Recognizing: The Judgment of Previous Occurrence." *Psychol. Rev.* **87** (1980): 252–271.

71. Marsolek, C. J., S. M. Kosslyn, and L. R. Squire. "Form-Specific Visual Priming in the Right Cerebral Hemisphere." *J. Exp. Psychol. Learn. Mem. Cog.* **18** (1992): 492–508.

72. Martone, M., N. Butters, and P. Payne. "Dissociations Between Skill Learning and Verbal Recognition in Amnesia and Dementia." *Arch. Neurol.* **41** (1984): 965–970.

73. McDougall, W. *An Outline of Psychology.* London: Methuen, 1923.

74. McKee, R. D., and L. R. Squire. "On the Development of Declarative Memory." *J. Exp. Psychol. Learn. Mem. Cog.* **19** (1993): 397–404.

75. Medin, D. L., and M. M. Schaffer. "Context Theory of Classification Learning." *Psychol. Rev.* **85** (1978): 207–238.

76. Milner, B., M. Petrides, and M. L. Smith. "Frontal Lobes and the Temporal Organization of Memory." *Human Neurobiol.* **4** (1985): 137–142.

77. Mishkin, M. "A Memory System in the Monkey." *Phils. Roy. Soc. Lond. [Biol.]* **298** (1982): 85–92.

78. Moscovitch, M., G. Winocur, and D. McLachlan. "Memory as Assessed by Recognition and Reading Time in Normal and Memory-Impaired People With Alzheimer's Disease and Other Neurological Disorders." *J. Exp. Psychol. Gen.* **115** (1986): 331–347.

79. Moss, M., H. Mahut, and S. Zola-Morgan. "Concurrent Discrimination Learning of Monkeys after Hippocampal, Entorhinal, or Fornix Lesions." *J. Neurosci.* **3** (1981): 227–240.

80. Murray, E. A., and M. Mishkin. "Severe Tactual as Well as Visual Memory Deficits Following Combined Removal of the Amygdala and Hippocampus in Monkeys." *J. Neurosci.* **4** (1984): 2565–2580.

81. Musen, G., A. P. Shimamura, and L. R. Squire. "Intact Text-Specific Reading Skill in Amnesia." *J. Exp. Psychol. Learn. Mem. Cog.* **6** (1990): 1068–1076.

82. Musen, G., and L. R. Squire. "Normal Acquisition of Novel Verbal Information in Amnesia." *J. Exp. Psychol. Learn. Mem. Cog.* **17** (1991): 1095–1104.

83. Musen, G., and L. R. Squire. "Nonverbal Priming in Amnesia." *Mem. Cog.* **20** (1992): 442–448.

84. Musen, G., and L. R. Squire. "Implicit Learning of Color-Word Associations Using a Stroop Paradigm." *J. Exp. Psychol. Learn. Mem. Cog.* **19** (1993): 789–798.

85. Musen, G., and L. R. Squire. "On the Implicit Learning of Novel Associations by Amnesic Patients and Normal Subjects." *Neuropsychologia* **7** (1993): 119–135.

86. Neely, J. H., and D. G. Payne. "A Direct Comparison of Recognition Failure Rates for Recallable Names in Episodic and Semantic Memory Tests." *Mem. Cog.* **11** (1983): 161–171.

87. Nissen, M. J., and P. Bullemer. "Attentional Requirements of Learning: Evidence from Performance Measures." *Cog. Psychol.* **19** (1987): 1–32.

88. Ostergaard, A. L. "Episodic, Semantic, and Procedural Memory in a Case of Amnesia at an Early Age." *Neuropsychologia* **25** (1987): 341–357.

89. Overman, W. H., G. Ormsby, and M. Mishkin. "Picture Recognition vs. Picture Discrimination Learning in Monkeys with Medial Temporal Removals." *Exp. Brain Res.* **79** (1991): 18–24.

90. Packard, M. G., R. Hirsh, and N. M. White. "Differential Effects of Fornix and Caudate Nucleus Lesions on Two Radial Maze Tasks: Evidence For Multiple Memory Systems." *J. Neurosci.* **9** (1989): 1465–1472.

91. Parkin, A. J. and B. M. Walter. "Recollective Experience, Normal Aging, and Frontal Dysfunction." *Psychol. Aging* **7** (1992): 290–298.

92. Perruchet, P., and C. Pacteau. "Synthetic Grammar Learning: Implicit Rule Abstraction or Explicit Fragmentary Knowledge?" *J. Exp. Psychol. Gen.* **119** (1990): 264–275.

93. Perruchet, P., and M. Amorim. "Conscious Knowledge and Changes in Performance in Sequence Learning: Evidence Against Dissociation." *J. Exp. Psychol. Learn. Mem. Cog.* **18** (1992): 785–800.

94. Phillips, R. R., B. L. Malamut, J. Bachevalier, and M. Mishkin. "Dissociation of the Effects of Inferior Temporal and Limbic Lesions on Object Discrimination Learning with 24-h Intertrial Intervals." *Behav. Brain Res.* **27** (1988): 99–107.

95. Posner, M. I., and S. W. Keele. "On the Genesis of Abstract Ideas." *J. Exp. Psychol.* **77** (1968): 353–363.

96. Reber, A. S. "Implicit Learning and Tacit Knowledge." *J. Exp. Psychol. Gen.* **118** (1989): 219–235.

97. Ribot, T. *Les Maladies de la Memoire [English translation: Diseases of Memory].* New York: Appleton-Century-Crofts, 1881.

98. Ringo, J. L. "Memory Decays at the Same Rate in Macaques With and Without Brain Lesions When Expressed in *d'* or Arcsine Terms." *Behav. Brain Res.* (1991): 123–134:

99. Roediger, H. L., III, and T. A. Blaxton. "Effects of Varying Modality, Surface Features, and Retention Interval on Priming in Word-Fragment Completion." *Mem. Cog.* **15** (1987): 379–388.

100. Rosch, E. H. 1973. "On the Internal Structure of Perceptual and Semantic Categories." In *Cognitive Development and the Acquisition of Language*, edited by T. E. Moore, 111–144. New York: Academic Press, 1973.

101. Sanderson, P. M. "Verbalizable Knowledge and Skilled Task Performance: Association, Dissociation, and Mental Models." *J. Exp. Psychol. Learn. Mem. Cog.* **15** (1989): 729–747.

102. Saunders, R. C., and L. Weiskrantz. "The Effects of Fornix Transection and Combined Fornix Transection, Mammillary Body Lesions and Hippocampal Ablations on Object-Pair Association Memory in the Rhesus Monkey." *Behav. Brain Res.* **35** (1989): 85–94.

103. Schacter, D. L., J. L. Harbluk, and D. R. McLachlan. "Retrieval Without Recollection: An Experimental Analysis of Source Amnesia." *J. Verbal Learn. Verbal Behav.* **23** (1984): 593–611.

104. Schacter, D. L., and P. Graf. "Preserved Learning in Amnesic Patients: Perspectives from Research on Direct Priming." *J. Clin. Exp. Neuropsy.* **6** (1986): 727–743.

105. Schacter, D. L. "Implicit Memory: History and Current Status." *J. Exp. Psychol. Learn. Mem. Cog.* **13** (1987): 501–518.

106. Schacter, D. L., L. A. Cooper, M. Tharan, and A. B. Rubens. "Preserved Priming of Novel Objects in Patients with Memory Disorders." *J. Cog. Neurosci.* **3** (1991): 118–131.

107. Schacter, D. L., and B. Church. "Auditory Priming: Implicit and Explicit Memory for Words and Voices." *J. Exp. Psychol. Learn. Mem. Cog.* **18** (1992): 915–930.

108. Schacter, D. L., C.-Y. Chiu, and K. N. Ochsner. "Implicit Memory: A Selective Review." *Ann. Rev. Neurosci.* **16** (1993): 159–182.

109. Schacter, D. L., and E. Tulving. *Memory Systems*. Cambridge, MA: MIT Press, 1994.

110. Scoville, W. B., and B. Milner. "Loss of Recent Memory After Bilateral Hippocampal Lesions." *J. Neurol., Neurosurg. Psychia.* **20** (1957): 11–21.

111. Servan-Schreiber, E., and J. R. Anderson. "Learning Artificial Grammars with Competitive Chunking." *J. Exp. Psychol. Learn. Mem. Cog.* **16** (1990): 592–608.

112. Shanks, D. R. "Categorization by a Connectionist Network." *J. Exp. Psychol. Learn. Mem. Cog.* **17** (1991): 433–443.

113. Sherry, D. F., and D. L. Schacter. "The Evolution of Multiple Memory Systems." *Psychol. Rev.* **94** (1987): 439–454.

114. Shimamura, A. P., and L. R. Squire. "A Neuropsychological Study of Fact Memory and Source Amnesia." *J. Exp. Psychol. Learn. Mem. Cog.* **13** (1987): 464–473.

115. Shimamura, A. P., and L. R. Squire. "Long-Term Memory in Amnesia: Cued Recall, Recognition Memory, and Confidence Ratings." *J. Exp. Psychol. Learn. Mem. Cog.* **14** (1988): 763–770.

116. Shimamura, A. P., and L. R. Squire. "Impaired Priming of New Associations in Amnesia." *J. Exp. Psychol. Learn. Mem. Cog.* **15** (1989): 721–728.

117. Shimamura, A. P., J. S. Janowsky, and L. R. Squire. "Memory for the Temporal Order of Events in Patients with Frontal Lobe Lesions and Amnesic Patients." *Neuropsychologia* **28** (1990): 803–814.

118. Shimamura, A. P., J. S. Janowsky, and L. R. Squire. "What is the Role of Frontal Lobe Damage in Memory Disorders?" In *Frontal Lobe Functioning and Dysfunction*, edited by H. D. Levin, H. M. Eisenberg, and A. L. Benton, 173–195. New York: Oxford University Press, 1991.

119. Shimamura, A. P., and L. R. Squire. "The Relationship Between Fact and Source Memory: Findings from Amnesic Patients and Normal Subjects." *Psychobiology* **19** (1991): 1–10.

120. Smith, M., and M. Oscar-Berman. "Repetition Priming of Words and Pseudowords in Divided Attention in Amnesia." *J. Exp. Psychol. Learn. Mem. Cog.* **16** (1990): 1033–1042.

121. Smith, M. E. "Neurophysiological Manifestations of Recollective Experience During Recognition Memory Judgments." *J. Cog. Neurosci.* **5** (1993): 1–13.

122. Squire, L. R., A. Shimamura, and P. Graf. "Independence of Recognition Memory and Priming Effects: A Neuropsychological Analysis." *J. Exp. Psychol. Learn. Mem. Cog.* **11** (1985): 37–44.

123. Squire, L. R. *Memory and Brain.* New York: Oxford University Press, 1987.

124. Squire, L. R., F. Haist, and A. P. Shimamura. "The Neurology of Memory: Quantitative Assessment of Retrograde Amnesia in Two Groups of Amnesic Patients." *J. Neurosci.* **9** (1989): 828–839.

125. Squire, L. R., and M. Frambach. "Cognitive Skill Learning in Amnesia." *Psychobiology* **18** (1990): 109–117.

126. Squire, L. R., and S. Zola-Morgan. "The Medial Temporal Lobe Memory System." *Science* **253** (1991): 1380–1386.

127. Squire, L. R. "Memory and the Hippocampus: A Synthesis from Findings with Rats, Monkeys, and Humans." *Psych. Rev.* **99** (1992): 143–145.

128. Squire, L. R., and R. McKee. "The Influence of Prior Events on Cognitive Judgments in Amnesia." *J. Exp. Psychol. Learn. Mem. Cog.* **18** (1992): 106–115.

129. Squire, L. R., J. G. Ojemann, F. M. Miezin, S. E. Petersen, T. O. Videen, and M. E. Raichle. "Activation of The Hippocampus in Normal Humans: A Functional Anatomical Study of Memory." *Proc. Natl. Acad. Sci. USA* **89** (1992): 1837–1841.

130. Squire, L. R., B. Knowlton, and G. Musen. "The Structure and Organization of Memory." *Ann. Rev. Psychol.* **44** (1993): 453–495.

131. Squire, L. R., and R. McKee. "Declarative and Nondeclarative Memory in Opposition: When Prior Events Influence Amnesic Patients more than Normal Subjects." *Mem. Cog.* **21** (1993): 424–430.

132. Srinivas, K., and H. L. Roediger. "Classifying Implicit Memory Tests: Category Association and Anagram Solution." *J. Mem. Lang.* **29** (1990): 389–412.

133. Thompson, R. F. "Neural Mechanisms of Classical Conditioning in Mammals." In *Behavioural and Neural Aspects of Learning and Memory*, edited by J. R. Krebs and G. Horn. Oxford: Clarendon Press, 1990.

134. Tolman, E. C. "There is More Than One Kind of Learning." *Psychol. Rev.* **56** (1949): 144–155.

135. Tulving, E., D. L. Schacter, and H. A. Stark. "Priming Effects in Word-Fragment Completion are Independent of Recognition Memory." *J. Exp. Psychol. Learn. Mem. Cog.* **8** (1982): 336–342.

136. Tulving, E. *Elements of Episodic Memory.* Cambridge: Oxford University Press, 1983.

137. Tulving, E. "How Many Memory Systems are There?" *Amer. Psychol.* **40** (1985): 385–398.

138. Tulving, E., D. L. Schacter, D. McLachlan, and M. Moscovitch. "Priming of Semantic Autobiographical Knowledge: A Case Study of Retrograde Amnesia." *Brain Cog.* **8** (1988): 3–20.

139. Tulving, E., C. A. G. Hayman, and C. A. MacDonald. "Long-Lasting Perceptual Priming and Semantic Learning in Amnesia: A Case Experiment." *J. Exp. Psychol. Lean. Mem. Cog.* **17** (1991): 595–617.

140. Tulving, E., personal communication, May, 1993.

141. Vokey, J. R., and L. R. Brooks. "The Salience of Item Knowledge in Learning Artificial Grammars." *J. Exp. Psychol. Learn. Mem. Cog.* **18** (1992): 328–344.

142. Wang, J., T. Aigner, and M. Mishkin. "Effects of Neostriatal Lesions on Visual Habit Formation in Rhesus Monkeys." *Soc. Neurosci. Abs.* **16** (1990): 617.

143. Warrington, E. K., and L. Weiskrantz. "Amnesia: A Disconnection Syndrome." *Neuropsychologia* **20** (1982): 233–248.

144. Weiskrantz, L. "Problems of Learning and Memory: One or Multiple Memory Systems?" *Phils. Roy. Soc. Lond. [Biol.]* **329** (1990): 99–108.

145. Willingham, D. B., M. J. Nissen, and P. Bullemer. "On the Development of Procedural Knowledge." *J. Exp. Psychol. Learn. Mem. Cog.* **15** (1989): 1047–1060.

146. Winocur, G. "Anterograde and Retrograde Amnesia in Rats with Dorsal Hippocampal or Dorosomedial Thalamic Lesions." *Behav. Brain Res.* **38** (1990): 145.

147. Wright, A. A., H. C. Santiago, S. F. Sands, D. F. Kendrick, and R. G. Cook. "Memory Processing of Serial Lists by Pigeons, Monkeys and People." *Science* **229** (1985): 287–289.

148. Zola-Morgan, S., L. R. Squire, and D. G. Amaral. "Human Amnesia and the Medial Temporal Region: Enduring Memory Impairment Following a Bilateral Lesion Limited to Field CA1 of the Hippocampus." *J. Neurosci.* **6** (1986): 2950–2967.

149. Zola-Morgan, S., L. R. Squire, and D. G. Amaral. "Lesions of the Amygdala that Spare Adjacent Cortical Regions Do not Impair Memory or Excerbate the Impairment Following Lesions of the Hippocampal Formation." *J. Neurosci.* **9** (1989): 1922–1936.

150. Zola-Morgan, S., and L. R. Squire. "The Primate Hippocampal Formation: Evidence for a Time-Limited Role in Memory Storage." *Science* **250** (1990): 288–290.

151. Zola-Morgan, S., and L. R. Squire. "Neuroanatomy of Memory." *Ann. Rev. Neurosci.* **16** (1993): 547–563.

152. Zola-Morgan, S., L. R. Squire, R. P. Clower, and N. L. Rempel. "Damage to the Perirhinal Cortex Exacerbates Memory Impairment Following Lesions to the Hippocampal Formation." *J. Neurosci.* **13** (1993): 251–265.

153. Zola-Morgan, S., L. R. Squire, and S. Ramus. "Severity of Memory Impairment in Monkeys as a Function of Locus and Extent of Damage within the Medial Temporal Lobe Memory System." *Hippocampus* (1993): in press.

Patricia Smith Churchland
University of California, San Diego Salk Institute, La Jolla, CA

Can Neurobiology Teach Us Anything About Consciousness?

1. INTRODUCTION

Human nervous systems display an impressive roster of complex capacities, including the following: perceiving, learning and remembering, planning, deciding, performing actions, as well as the capacities to be awake, fall asleep, dream, pay attention, and be aware. Although neuroscience has advanced spectacularly in this century, we still do not understand in satisfying detail how any capacity in the list emerges from networks of neurons.[1] We do not completely understand how humans can be conscious, but neither do we understand how they can walk, run, climb trees, or pole vault. Nor, when one stands back from it all, is awareness intrinsically more mysterious than motor control. Balanced against the disappointment that full understanding eludes us still, is cautious optimism, based chiefly on the nature of the progress behind us. For cognitive neuroscience has already passed

---

[1] See our discussion in *The Computational Brain*, by Churchland and Sejnowski.[17]

well beyond what skeptical philosophers once considered possible, and continuing progress seems likely.

In assuming that neuroscience can reveal the physical mechanisms subserving psychological functions, I am assuming that it is indeed the brain that performs those functions—that capacities of the human mind are, in fact, capacities of the human brain. This assumption and its corollary rejection of Cartesian souls or spirits or "spooky stuff" existing separately from the brain is no whimsy. On the contrary, it is a highly probable hypothesis, based on evidence currently available from physics, chemistry, neuroscience, and evolutionary biology. In saying that physicalism is an hypothesis, I mean to emphasize its status as an empirical matter. I do not assume that it is a question of conceptual analysis, a priori insight, or religious faith, though I appreciate that not all philosophers are at one with me on this point.[2]

Additionally, I am convinced that the right strategy for understanding psychological capacities is essentially reductionist, by which I mean, broadly, that understanding the neurobiological mechanisms is not a frill but a necessity. Whether science will finally succeed in reducing psychological phenomena to neurobiological phenomena is, needless to say, yet another empirical question. Adopting the reductionist strategy means trying to explain the macro levels (psychological properties) in terms of micro levels (neural network properties).

The fundamental rationale behind this research strategy is straightforward: if you want to understand how a thing works, you need to understand not only its behavioral profile, but also its basic components and how they are organized to constitute a system. If you do not have the engineering designs available for reference, you resort to reverse engineering—the tactic of taking apart a device to see how it works.[3] Insofar as I am trying to discover macro-to-micro explanations, I am a reductionist. Because many philosophers who agree with me on the brain-based nature of the soul nonetheless rail against reductionism as ridiculous if not downright pitiful, it may behoove me to begin by explaining briefly what I do and, most emphatically, do *not* mean by a reductionist research strategy.[4]

Clearing away the "negatives" first, may I say that I do *not* mean that a reductionist research strategy implies that a *purely bottom-up strategy* should be adopted. So far as I can tell, no one in neuroscience thinks that the way to understand the nervous system is first to understanding everything about the basic molecules, then everything about every neuron and every synapse, and to continue ponderously thus to ascend the various levels of organization until, at long last, one arrives at the uppermost level—psychological processes (see Figure 1). Nor is there anything in the history of science that says a research strategy is reductionist only if it is

[2] For concordant opinions, see also Francis Crick[18]; Paul Churchland[16]; Daniel Dennett[24]; Owen Flanagan[28]; William G. Lycan[39]; and John Searle.

[3] As P. S. Churchland and T. J. Sejnowski have argued.[16]

[4] For an outstanding discussion of reductionism that includes many of the complexities I am not worrying too much about here, see Schaffner.[45]

purely bottom-up. That characterization is straw through and through. The research behind the classical reductionist successes—explanation of thermodynamics in terms of statistical mechanics; of optics in terms of electromagnetic radiation; of hereditary transmission in terms of DNA—certainly did not conform to any purely bottom-up research directive.

So far as neuroscience and psychology are concerned, my view is simply that it would be wisest to conduct research on many levels simultaneously, from the molecular, through to networks, systems, brain areas, and, of course, behavior.

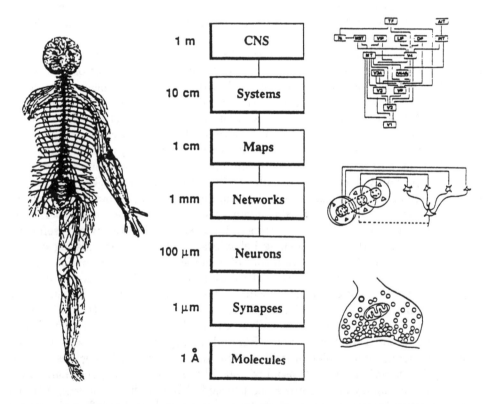

FIGURE 1   Schematic illustration of levels of organization in the nervous system. The spatial scales at which anatomical organization can be identified varies over many orders of magnitude. The icon to the left depicts the "neuron man," showing the brain, spinal cord, and peripheral nerves. The icons to the right represent structures at distinct levels: (top) a subset of visual areas in visual cortex; (middle) a network model proposing how ganglion cells could be connected to "simple" cells in visual cortex; and (bottom) a chemical synapse. (From Churchland and Sejnowski.[17])

Here, as elsewhere in science, hypotheses at various levels can *coevolve* as they correct and inform one another.[11] Neuroscientists would be silly to make a point of ignoring psychological data, for example, just as psychologists would be silly to make a point of ignoring all neurobiological data.

Second, by "reductionist research strategy" I do not mean that there is something disreputable, unscientific, or otherwise unsavory about high-level descriptions or capacities *per se*. It seems fairly obvious, to take a simple example, that certain rhythmic properties in nervous systems are network properties resulting from the individual membrane traits of various neuron types in the network, together with the way the set of neurons interact. Recognition that something is the face of Arafat, for another example, almost certainly emerges from the responsivity profiles of the neurons in the network plus the ways in which those neurons interact. "Emergence" in this context is entirely nonspooky and respectable, meaning, to a first approximation, "property of the network." Determining precisely what the network property is, for some particular feat, will naturally take quite a lot of experimental effort. Moreover, given that neuronal behavior is highly nonlinear, the network properties are *never* a simple "sum of the parts." They are some function—some *complicated* function—of the properties of the parts. High-level capacities clearly exist, and high-level descriptions are therefore needed to specify them.

"Eliminative materialism" refers to the hypothesis that (1) materialism is most probably true, and also (2) many traditional aspects of explanation of human behavior are probably not adequate to the reality of the etiology of behavior.[5] The standard analogy here is that just as "caloric fluid" was useful but fundamentally mistaken in understanding thermal phenomena (conduction, convection, radiation) so some psychological categories currently invoked may be somewhat useful but fundamentally mistaken in fathoming behavioral etiology. Other existing characterizations of capacities may have a core of adequacy but undergo major redrawings, in something like the way Mendel's notion of "factor" came to be modified by genetics into the notion of "gene" which itself was modified and deepened with the development of molecular biology. Some categories such as "attitude" are extremely vague and might be replaced altogether; others, such as "is sleeping" have already undergone a fractionation as EEG and neurophysiological research has revealed important brain differences in various stages of sleep. Categories such as "memory," "attention," and "reasoning" are likewise undergoing revision, as experimental psychology and neuroscience proceed.[6] It remains to be seen whether there is a neurobiological reality to sustain notions such as "belief" and "desire" as articulated by modern philosophers such as Fodor and Searle,[46] though Paul

[5]See Paul Churchland's characterization and defense of this view, reprinted in P. M. Churchland.[16]

[6]See P. S. Churchland.[11]

Churchland and I have argued that revision here too is most probable.[7] The revisionary prediction, too, is an empirical hypothesis, and one for which empirical support already exists.[16,17]

The possibility of nontrivial revision and even replacement of existing high-level descriptions by "neurobiologically harmonious" high-level categories is the crux of what makes eliminative materialism *eliminative*.[8] By "neurobiologically harmonious" categories, I mean those that permit coherent, integrated explanations from the whole brain on down through neural systems, big networks, micronets, and neurons. Only the strawman is so foolish as to claim that there are no high-level capacities, that there are no high-level phenomena.[10] In its general aspect, my point here merely reflects this fact: in a profoundly important sense, we do not understand exactly what, at its higher levels, the brain really does. Accordingly, it is practical to earmark even our fondest intuitions about mind/brain function as revisable hypotheses rather than as transcendental absolutes or introspectively given certainties. Acknowledgment of such revisability makes an enormous difference in how we conduct psychological and neurobiological experiments, and in how we interpret the results.

2. NAYSAYING THE NEUROBIOLOGICAL GOAL

Over the last several decades, a number of philosophers have expressed reservations concerning the reductionist research goal of discovering the neurobiological mechanisms for psychological capacities, including the capacity to be conscious. Consequently, it may be useful to consider the basis for some of these reservations in order to determine whether they justify abandoning the goal, or whether they should dampen our hopes about what might be discovered about the mind/brain. I shall here consider five main classes of objection. As a concession to brevity, my responses shall be ruthlessly succinct, details being sacrificed for the sake of the main gist.

2.1 THE GOAL IS ABSURD (INCOHERENT)

One set of reasons for dooming the reductionist research strategy is summed up thus: "I simply cannot imagine that seeing blue or the feeling of pain, for example, could consist in some pattern of activity of neurons in the brain," or, more bluntly, "I cannot imagine how you can get awareness out of meat." There is sometimes considerable filler between the "it's unimaginable" premise and the "it's impossible"

[7]Op. cit.

[8]Or, as we have preferred but decided not to say, "what makes revisionary materialism revisionary."[8,12] For a related but somewhat different picture, see Bickle.[3]

conclusion, but, so far as I can tell, the filler is typically dust which cloaks the fallacious core of the argument.[40]

Given how little in detail we currently understand about how the human brain "en-neurons" any of its diverse capacities, it is altogether predictable that we should have difficulty imagining the neural mechanisms. When the human scientific community was comparably ignorant of such matters as valence, electron shells, and so forth, natural philosophers could not imagine how you could explain the malleability of metals, the magnetizability of iron, and the rust resistance of gold, in terms of underlying components and their organization. Until the advent of molecular biology, many people thought it was unimaginable, and hence impossible, that to be a living thing could consist in a particular organization of "dead" molecules. "I cannot imagine," said the vitalists, "how you could get *life* out of *dead* stuff."

From the vantage point of considerable ignorance, failure to imagine some possibility is only that: a failure of imagination—one psychological capacity amongst others. It does not betoken any metaphysical limitations on what we can come to understand, and it cannot predict anything significant about the future of scientific research. After reflecting on the awesome complexity of the problem of thermoregulation in homeotherms such as ourselves, I find I cannot imagine how brains control body temperature under diverse conditions. I suspect, however, that this is a relatively uninteresting psychological fact about me, reflecting merely my current state of ignorance. It is not an interesting metaphysical fact about the universe nor even an epistemological fact about the limits of scientific knowledge.

A variation of the "cannot imagine" proposal is expressed as "we can never, never know...," or "it is impossible to ever understand...," or "it is forever beyond science to show that...." The idea here is that something's being impossible to conceive says something decisive about its empirical or logical impossibility. I am not insisting that such proposals are never relevant. Sometimes they may be. But they are surprisingly high handed when science is in the very early stages of studying a phenomenon.

The sobering point here is that assorted "*a priori* certainties" have, in the course of history, turned out to be empirical duds, however obvious and heartfelt in their heyday. The impossibility that space is non-Euclidean, the impossibility that in real space, parallel lines should converge, the impossibility of having good evidence that some events are undetermined, or that someone is now dreaming, or that the universe had a beginning—each slipped its logical noose as we came to a deeper understanding of how things are. If we have learned anything from the many counterintuitive discoveries in science, it is that our intuitions can be wrong. Our intuitions about ourselves and how we work may also be quite wrong. There is no basis in evolutionary theory, mathematics, or anything else, for assuming that prescientific conceptions are essentially scientifically adequate conceptions.

A third variation on this "nay, nay, never" theme draws conclusions about how the *world must actually be*, based on *linguistic properties* of certain central categories in current use to describe the world. Permit me to give a boiled down instance: "the category "mental" is remote in meaning—means something completely different—from the category physical. It is absurd therefore to talk of the brain seeing or feeling, just as it is absurd to talk of the mind having neurotransmitters or conducting current." Allegedly, this categorial absurdity undercuts the very possibility that science could discover that feeling pain is an activity in neurons in the brain. The epithet "category error" is sometimes considered sufficient to reveal the naked nonsense of reductionism.

Much has already been said on this matter elsewhere,[9] and I shall bypass a lengthy discussion of philosophy of language with three brief points. (1) It is rather far-fetched to suppose that intuitions in the philosophy of language can be a reliable guide to what science can and cannot discover about the nature of the universe. (2) Meanings *change* as science makes discoveries about what some macro phenomenon *is* in terms of its composition and the dynamics of the underlying structure. (3) Scientists are unlikely to halt their research when informed that their hypotheses and theories "sound funny" relative to current usage. More likely, they will say this: "The theory might sound funny to you, but let me teach the background science that makes us think the theory is true. Then it will sound less funny." It may be noted that it sounded funny to Copernicus' contemporaries to say the Earth is a planet and moves; it sounded funny to say that heat is molecular motion or that physical space is non-Euclidean or that there is no absolute "downness." And so forth.

That a scientifically plausible theory sounds funny is a criterion only of its not having become common song, not of its being wrong. Scientific discoveries that a certain macro phenomenon is a complex result of the micro structure and its dynamics are typically surprising and typically sound funny—at first. Obviously none of this is positive evidence that we can achieve a reduction of psychological phenomena to neurobiological phenomena. It says only that sounding funny does not signify anything, one way or the other.

2.2 THE GOAL IS INCONSISTENT WITH "MULTIPLE REALIZABILITY"

The core of this objection is that if a macro phenomenon can be the outcome of more than one mechanism (organization and dynamics of components), then it cannot be identified with any one mechanism, and hence the reduction of the macrophenonomenon to *the* (singular) underlying micro phenomenon is impossible. This objection seems to me totally uninteresting to science. Again, permit me to ignore important details and merely to summarize the main thrust of the replies.

1. Explanations, and therefore reductions, are domain relative. In biology, it may be fruitful first to limn the general principles explaining some phenomenon seen in diverse species, and then figure out how to account for the interspecies differences, and then, if desirable, how to account for differences across individuals

[9] See, for example, Feyerabend.[27]

within a given species. Thus the general principles of how hearts or stomachs work are figured out, perhaps based on studies of a single species, and particularities can be resolved thereafter. Frog hearts, macaque hearts, and human hearts work in essentially the same general way, but there are also significant differences, apart from size, that call for comparative analyses. Consider other examples: (a) from the general solution to the copying problem that emerged from the discovery of the fundamental structure of DNA, it was possible to undertake explorations of how differences in DNA could explain certain differences in the phenotype; and (b) from the general solution to the problem of how neurons send and receive signals, it was possible to launch detailed exploration into the differences in responsivity profiles of distinct classes of neuron (see also Owen Flanagan[29]).

2. Once the mechanism for some biological process has been discovered, it may be possible to invent devices to mimic those processes. Nevertheless, invention of the technology for artificial hearts or artificial kidneys does not obliterate the explanatory progress on actual hearts and actual kidneys; it does not gainsay the reductive accomplishment. Again, the possibility that hereditary material of a kind different from DNA might be found in things elsewhere in the universe does not affect the basic scaffolding of a reduction on this planet. Science would have been much the poorer if Crick and Watson had abandoned their project because of the abstract possibility of Martian hereditary material or artificial hereditary material. In fact, we do know the crux of the copying mechanism *on Earth*—namely, DNA—and we do know quite a lot about how it does its job. Similarly, the engineering of artificial neurons and artificial neural nets (ANNs) facilitates and is facilitated by neurobiological approaches to how real neurons work; the engineering undertakings do not mean the search for the basic principles of nervous system function is misguided.

3. There are always questions remaining to be answered in science, and hence coming to grasp the general go of a mechanism, such as the discovery of base pairing in DNA, ought not be mistaken for the utopian ideal of a complete reduction—a complete explanation. Discoveries about the general go of something typically raise hosts of questions about the *detailed* go of it, and then about the details of the *details*. To signal the incompleteness of explanations, perhaps we should eschew the expression "reduction" in favor of "reductive contact." Hence we should say the aim of neuroscience is to make rich reductive contact with psychology as the two broad disciplines coevolve. I have experimented with this recommendation myself, and although some philosophers warm to it, scientists find it quaintly pedantic. In any case, "reductive contact" between molecular biology and macrobiology has become steadily richer since 1953, though many questions remain. Reductive contact between psychology and neuroscience has also become richer, especially in the last decade, though it is fair to say that by and large the basic principles of how the brain works are poorly understood.

4. What, precisely, are supposed to be the progammatic sequelae to the multiple realizability argument? Is it that neuroscience is *irrelevant* to understanding

the nature of the human mind? Obviously not. That neuroscience is *not necessary* to understand the human mind? One cannot, certainly, deny that it is remarkably useful. Consider the discoveries concerning sleep, wakeness, and dreaming; the discoveries concerning split brains, humans with focal brain lesions, the neurophysiology and neuroanatomy of the visual system, and so on. Is it perhaps that we should not get our hopes up too high? What, precisely, is "too high" here? Is it the hope that we shall discover the general principles of how the brain works? Why is that too high a hope?

2.3 THE BRAIN CAUSES CONSCIOUSNESS

Naysaying the reductionist goal while keeping dualism at arm's length is a manoeuvre requiring great delicacy. John Searle's strategy is to say that although the brain *causes* conscious states, any identification of conscious states with brain activities is unsound. Traditionally, it has been opined that the best the reductionist can hope for are *correlations* between subjective states and brain states, and although correlations can be evidence for causality, they are not evidence for identity. Searle has tried to bolster that objection by saying that whereas α/β identifications elsewhere in science reveal the reality behind the appearance, in the case of awareness, the reality and the appearance are inseparable—there is no reality to awareness except what is present in awareness. There is, therefore, no reduction to be had.

Synoptically, here is why Searle's manoeuvre is unconvincing: he fails to appreciate why scientists opt for identifications when they do. Depending on the data, cross-level identifications to the effect that α is β may be less troublesome and more comprehensible scientifically than supposing thing α causes separate thing β. This is best seen by example.[10]

Science, as we know it, says electrical current in a wire is not caused by moving electrons; it *is* moving electrons. Genes are not caused by chunks of base pairs in DNA; they *are* chunks of base pairs (albeit sometimes distributed chunks). Temperature is not caused by mean molecular kinetic energy; it *is* mean molecular kinetic energy. Reflect for a moment on the inventiveness required to generate explanations that maintain the *nonidentity* and causal dependency of (a) electric current and moving electrons, (b) genes and chunks of DNA, and (c) heat and molecular motion. Unacquainted with the relevant convergent data and explanatory successes, one may suppose this is not so difficult. Enter Betty Crocker.

In her microwave oven cookbook, Betty Crocker offers to explain how a microwave oven works. She says that when you turn the oven on, the microwaves excite the water molecules in the food, causing them to move faster and faster. Does she, as any high school science teacher knows she should, end the explanation

[10] In the following discussion, the ideas are mostly owed to Paul Churchland.[8] For his discussion, see "Betty Crocker's Theory of the Mind: A Review of John Searle's *The Rediscovery of the Mind." The London Review of Books*, May, 1994.

here, perhaps noting, "increased temperature just is increased kinetic energy of the constituent molecules?" She does not. She goes on to explain that because the molecules move faster, they bump into each other more often, which increases the friction between molecules, and, as we all know, friction causes heat. *Betty Crocker still thinks heat is something other than molecular KE; something caused by, but actually independent of, molecular motion.*[11] Why do scientists not think so, too?

Roughly, because explanations for heat phenomena—production by combustion, by the sun, and in chemical reactions; of conductivity, including conductivity in a vacuum, the variance in conductivity in distinct materials, etc.—are *vastly* simpler and more coherent on the assumption that heat is molecular energy of the constituent molecules. By contrast, trying to make the data fit with assumption that heat is some other thing *caused by* speeding up molecular motion is like trying to nail jelly to the wall.

If one is bound and determined to cleave to a caloric thermodynamics, one might, with heroic effort, pull it off for oneself, though converts are improbable. The cost, however, in coherence with the rest of scientific theory, not to mention with other observations, is extremely high. What would motivate paying that cost? Perhaps an iron-willed, written-in-blood resolve to maintain unsullied the intuition that heat "*is what it is and not another thing.*" In retrospect, and knowing what we now know, the idea that anyone would go to exorbitant lengths to defend the "heat intuition" seems rather a waste of time.

In the case at hand, I am predicting that explanatory power, coherence, and economy will favor the hypothesis that awareness just *is* some pattern of activity in neurons. I may turn out to be wrong. If I am wrong, it will not be because an introspectively based intuition is immutable, but because the science leads us in a different direction. If I am right, and certain patterns of brain activity *are* the reality behind the experience, this fact does not in and of itself change my experience and suddenly allow me (my brain) to view my brain as an MR scanner or a neurosurgeon might view it. I shall continue to have experiences in the regular old way, though in order to understand the neuronal reality of them, my brain needs to *have* lots of experiences and undergo lots of learning.

Finally, barring a jump to the dualist' s horse, the idea that there has to be a bedrock of subjective "appearance" on which reality/appearance discoveries must ultimately rest is faintly strange. It seems a bit like insisting that "down" cannot be relative to where one is in space; down is down. Or like insisting that time cannot be relative, that either two events happen at the same time or they don't, and that's that. Humans are products of evolution; nervous systems have evolved in

---

[11] Paul Churchland made this discovery in our kitchen about eight years ago. It seemed to us a bang-up case of someone's not really understanding the scientific explanation. Instead of thinking the thermodynamic theory through, Betty Crocker just clumsily grafts it onto on old conception as though the old conception needed no modification. Someone who thought electricity was caused by moving electrons would tell a comparable Betty Crocker story: "voltage forces the electrons to move through the wire, and as they do so, they cause static electricity to build up, and a sparks then jump from electron to electron, on down the wire."

the context of competition for survival—in the struggle to succeed in the four F's: feeding, fleeing, fighting, and reproduction. The brain's model of the external world enjoys improvement through appreciating various reality/appearance distinctions— in short, through common critical reason and through science. In the nature of things, it is quite likely that the brain's model of its internal world also allows for appearance/reality discoveries. The brain did not evolve to know the nature of the sun as it is known by a physicist, nor to know itself as it is known by a neurophysiologist. But, in the right circumstances, it can come to know them anyhow.[12]

2.4 BECAUSE CONSCIOUSNESS IS A VIRTUAL MACHINE

This is the view of D. C. Dennett.[24] Like Searle, Dennett is no dualist. Unlike Searle, who thinks that quite a lot, if not all, about consciousness can be discovered by neuroscience, Dennett has long been convinced that study of the brain itself— its physiology and anatomy—is largely a waste of time so far as understanding the nature of consciousness and cognition are concerned. Simplified, the crux of his idea is this: humans become conscious as they acquire language and learn to talk to themselves. What happens in this transformation is that a parallel machine (the neural networks of the brain) simulates a serial machine (operations are performed one at a time, in a sequence, according to rules, which may be recursive.)

By acquiring a language and then learning to speak silently to oneself, one allegedly creates a "consciousness virtual machine" in the brain. Dennett explains what this is by means of a pivotal analogy: it is like creating a virtual machine for simulating piloting a plane on your desktop computer by installing software, such as a *Flight Simulator*. Consciousness bears the same relation to the brain as the flight simulation bears to the events inside the computer. Dennett's methodological moral is unambiguous: just as we cannot hope to learn anything much about the flight simulator (its scope and limits, how it works) by studying the computer's innards while it is running *Flight Simulator* so we cannot hope to learn much about consciousness by studying the brain's innards while it is conscious. If one wants to know about *Flight Simulator* and its many properties, the best you can do is study its performance—in a sense, there really is not anything else to *Flight Simulator* than its performance. We find it fruitful in talking about *Flight Simulator* to say things like "its altimeter registers altitude," but this does not mean that there is something in my computer that really is high in the sky or something that measures how high it really is. Such talk is simply an economical, convenient way of making sense of the computer's screen performance when it is running *Flight Simulator* software.

Ditto (more or less) for consciousness. The brain is the hardware on which the consciousness "software" runs, and hence looking at the brain itself is not going

[12] See P. M. Churchland.[9]

to teach us much about the software itself. Even as it is a mistake to suppose the computer has a little runway hidden tucked inside that gets rolled out when I press a button, so it is a mistake to think the brain really does anything like fill in the blindspot or fill in during seeing "subjective motion" (as in a movie).[13] Dennett believes he has shown us that there really is not so much in the way of "inner experience" to be explained after all. As with *Flight Simulator*, if you want to know about consciousness and its properties, it is performance under a variety of conditions that needs to be studied. Based on the performance you can, of course, infer the various computational properties of the software. *And that is all there will be to explaining consciousness.* Consequently, the tools of experimental psychology will suffice. The details of neuroscience might tell us something about how the "software" runs on the brain; that won't tell us anything about the nature of consciousness, but only about how the brain runs "software." This, in capsule, is my understanding of the conviction that inspires Dennett to his book's title, *Consciousness Explained.*

How plausible is the Dennett story? My criticism here draws on work of Paul Churchland[15] and will focus mainly on this question: is it remotely reasonable that when we are conscious the parallel machine (the brain) is simulating a serial machine? As an archival preliminary, however, note that Dennett's package has been subjected to intense and careful analysis. First, his claim that acquisition of human language is a necessary condition of human consciousness has been repeatedly challenged and thoroughly criticized.[4,28] Endlessly it has been noted that this seems to imply that preverbal infants are not conscious; that other animals, such as chimpanzees and orangutans, are not conscious; that subjects with global aphasia or left hemispherectomies are not conscious. Briefly, Dennett's response is this: yes, indeed, nonverbal subjects are not aware in the way a fully verbal human is aware; e.g., they cannot think about whether interest rates will go down next month. Unfortunately, Dennett's response is tangential to the criticism. The issue is whether preverbal children and animals can be conscious of colors, sounds, smells, spatial extent, motion, being dizzy, feeling pain, etc. in rather like the way I am conscious of them.

Second, Dennett's according pre-eminent status to linguistic activity and his correlative "debunking" of sensory experiences (e.g., filling in), feelings, and nonlinguistic cognition generally, have been subject to a constant barrage of complaints.[14] Regrettably, I can give here only a highly truncated version of the long and sometimes convoluted debates between Dennett and various critics. The heart of the complaints is that Dennett wrongly assumes that performance is all that needs explaining—that explaining *reports of* conscious experience is tantamount to explaining conscious experience itself. Dennett's core response here has been to wave off the complainers as having failed properly to understand him, scolding them for being still in the grip of bad old conceptual habits implying homunculi, ghosts

---

[13] For a criticism of Dennett on filling in, see Churchland and Ramachandran.
[14] Ibid. Also John Searle's *The Rediscovery of the Mind.*[46]

in the machine, furtive Cartesianism, and kindred mistakes. Suffice it to say that Dennett's "if-you-disagree-you-have-misunderstood" stance, while conceivably true of some critics, does not appear true of all.

Is a "virtual serial machine" *needed* to get a one-after-another temporal ordering? Not at all. For example, it has been well known for at least eight years that neural nets with recurrent loops can yield temporal sequencing, and do so very economically and elegantly.[41,47,49] For a recent example, beautiful work in using "real-valued genetic algorithms to evolve continuous-time recurrent neural networks capable of sequential behavior and learning" has been done by Randall Beer and other sequencing work has been done by Michael Mozer. Clearly, sequencing tasks *per se* do not imply the existence of a simulated serial machine.[17]

Is a virtual serial machine needed to get rule-following behavior as seen in linguistic performance? Not at all. Again, as Elman and his colleagues have shown, recurrent neural nets can manage this very well.[25,30,33,42,43,44,48] Is a virtual serial machine needed to restrict a certain class of operations to one at a time? Not at all. First, a special class of operations could be the output of one network, albeit a widely distributed network. Second, they could be the output of a winner-take-all interaction between nets.[35] And there are lots of other architectures for accomplishing this. The motor system probably functions thus, but there is no reason to think it *simulates* a serial machine.[2,51]

Third, should we assume that consciousness involves only one operation at a time? Almost certainly not. Granting that the attentional capacity is much smaller than the extra-attentive capacity to represent,[50] why conclude that we can attend to only one thing at a time? When I look at a bowl of colored M&M's, can I see more than one M&M at once? Probably, and while humming "Eleanor Rigby." Fourth, is the serial machine simulation necessary in order to enable recursive properties, such that one can be self-aware (think about what one just said to oneself)? Not at all. Recurrent neural nets are powerful enough and complex enough to manage this very nicely. Indeed, recurrence probably is a key feature of various self-monitoring subsystems in the nervous system, including thermoregulation. *Is* there any rationale for saying, "when we are conscious the brain must be simulating a serial machine?" I see none. This does not entail that Dennett must be wrong, but only that there is no reason to think he is on the right track.

2.5 THE PROBLEM IS BEYOND OUR FEEBLE INTELLIGENCE

Initially, this claim appears to be a modest acknowledgment of our limitations.[40] In fact, it is a powerful prediction based not on solid evidence, but on profound ignorance. For all we can be sure now, the prediction might be correct, but equally, it might very well be false. How feeble is our intelligence? How difficult is the problem? How could you possibly know that solving the problem is beyond our reach, no matter how science and technology develop? Inasmuch as it is not known that the brain is more complicated than it is smart, giving up on the attempt to find out how it works would be disappointing. On the contrary, as long as experiments continue to produce results that contribute to our understanding, why not keep going?[15]

3. TRACKING DOWN THE NEURAL MECHANISMS OF CONSCIOUSNESS

3.1 FINDING A ROUTE IN

In neuroscience there are many data at higher levels relevant to consciousness. Blindsight, hemineglect, split brains, anosognosia (unawareness of deficit), for starters, are powerful constraints to guide theoretical reflection. Careful studies using scanning devices such as magnetic resonance imaging (MRI) and positron emission tomography (PET) have allowed us to link specific kinds of functional losses with particular brain regions.[20,21,22,26] This helps narrow the range of structures we consider selecting for preliminary micro exploration.

For example, the hippocampus might have seemed a likely candidate for a central role in consciousness because it is a region of tremendous convergence of fibres from diverse areas in the brain. We now know, however, that bilateral loss of the hippocampus, though it impairs the capacity to learn new things, does not entail loss of consciousness. At this stage, ruling something out is itself a valuable advance. We also know that certain brainstem structures such as the locus coereleus (LC) are indirectly necessary, but are not part of the mechanism for consciousness. LC does play a nonspecific role in arousal, but not a specific role in awareness of particular contents, such as awareness at a moment of the color of the morning sky rather than the sound of the lawn sprinklers. The data may be fascinating in its own right, but the question remains: how can we get from an array of intriguing data to genuine explanations of the basic mechanism? How can we get *started*?

In thinking about this problem, I have been greatly influenced by Francis Crick. His basic approach is straightforward: if we are going to solve the problem, we should treat it as a scientific problem to be tackled in much the way we tackle other difficult scientific problems. As with any scientific mystery, what we want is a revealing experimental entry. We want to find a thread which, when pulled, will unloose a whole lot else. To achieve that, we need to devise testable hypotheses that can connect macro effects with micro dynamics.

Boiled down, what we face is a constraint satisfaction problem: find psychological phenomena that (a) have been reasonably well studied by experimental psychology, (b) are circumscribed by lesion data from human patients as well as data from precise animal microlesions, (c) are known to be related to brain regions where good neuroanatomy and neurophysiology has been done, and (d) where we

---

[15] See Owen Flanagan's convincing and more detailed discussion of McGinn's naysaying.[28]

know quite a lot about connectivity to other brain regions. The working assumption is that if a person is aware of a stimulus, his brain will be different in some discoverable respect from the condition where he is awake and attentive but unaware of the stimulus. An auspicious strategy is to hunt down those differences, guided by data from lesion studies, PET scans, magnetoencephalograph (MEG) studies, and so forth. Discovery of those differences, in the context of neurobiological data generally, should aid discovering a theory of the mechanism.

The central idea is to generate a theory constrained by data at many levels of brain organization—sufficiently constrained so that it can be meaningfully tested. Ultimately a theory of consciousness will need to encompass a range of processes involved in awareness, including attention and short-term memory. Initially, however, it may target a subset, such as integration across space and across time. Whether the theory falls to falsifying evidence or whether it survives tough tests, we shall learn something. That is, either we shall have ruled out specific possibilities—a fine prize in the early stages of understanding—or we can go on deepen and develop the theory further—an even finer prize. In any case, the trick is to generate testable, meaty hypotheses rather than loose, frothy hypotheses susceptible only to experiments of fancy. The trick is to make some real progress.

3.2 VISUAL AWARENESS

What plausible candidates surface from applying the constraint satisfaction procedure? Interestingly, the choices are quite limited. Although metacognition, introspection, and awareness of emotions, for example, are indeed aspects of consciousness, either we do not have good lesion data to narrow the search space of relevant brain regions, or the supporting psychophysics is limited or both. Consequently, these processes are best put on the back burner for later study.

Visual awareness, by contrast, is a more promising candidate. In the case of vision, as Crick points out, there is a huge literature in visual psychophysics to draw upon, there is a rich literature of human and animal lesion studies, and relative to the rest of the brain, a lot is known about the neuroanatomy and neurophysiology of the visual system, at least in the monkey and the cat. Visual phenomena such as filling in, binocular rivalry, seeing motion, seeing stereoptic depth, and so forth might reward the search for the neurobiological differences between being aware and not being aware in the awake, attentive animal. This may get us started, and I do emphasize *started*.

THE CRICK HYPOTHESIS. Immersed in the rich context of multilevel detail, Crick has sketched an hypothesis concerning the neuronal structures that he conjectures make the salient differences, depending on whether the animal is or is not visually aware of the stimulus.[16] Integration of representations across spatially distributed neural networks—the unity in apperception, so to speak—is thought be accomplished by temporal "binding," namely, synchrony in the output responses of the relevant neurons. Very crudely, Crick's suggestion is that (1) for sensory awareness, such as visual awareness, the early cortices are pivotal (e.g., visual areas V2, V3, V5). This makes sense of lesion data, as well as recent PET data,[23,34] and single cell data.[38] (2) Within the early sensory cortical areas, pyramidal cells in layer 5 and possibly layer 6, play the key role.

What good is this idea? Part of its appeal is its foothold in basic structure. In biology, the solution to difficult problems about mechanism can be greatly facilitated by identification of critical structures. Crudely, if you know "what," it helps enormously in figuring out "how." On its own, the Crick hypothesis can be only a small piece of the puzzle. If we are lucky, however, it, or something like it, may be a *key* piece of the puzzle. This is not the time for a fuller discussion of this hypothesis. Suffice it to say that true or false, the Crick hypothesis provides a bold illustration of how to approach a problem so tricky that it is often scrapped as unapproachable.

THE LLINAS HYPOTHESIS. Another promising entry route is suggested by the differences—phenomenological and neurobiological—between sleep/dreaming/wakeness[17] (SDW) states.[18] This entry point is attractive first because there is the familiar and dramatic loss of awareness in deep sleep, which is recovered as we awake, and is probably present also during dreaming. The phenomenon is highly available in lots of different subjects and across many species. Second, MEG and EEG techniques reveal global brain features characteristic of different states. Human and animal lesion data are important, especially as they concern deficits in awareness during wakeness. Here again I note the significance of research on blindsight, hemineglect (tendency to be unaware of stimuli in various modalities on the left side of the body), simultanagnosia (inability to see several things simultaneously), anosognosia (unawareness of deficits such as paralysis), blindness denial, unawareness ofjargon aphasia (of not making sense), and so forth.

Third, we have learned a great deal from abnormalities in and manipulation of the SDW cycle and the link to specific brain properties. Fourth, some of the global changes in state in the SDW cycle seen by macro techniques have been

[16] This point is made by Crick and Koch[19] and by Crick.[18]

[17] Wakeness is a bit of a neologism. I prefer not to use "wakefulness" inasmuch as it connotes some difficulty in sleeping, or being easily awoken, as in "the baby was wakeful when he had a rash." I prefer it also to the phrase "being awake" which is both rather cumbersome and fails to preserve parallism with "dreaming," "sleeping." Clearly one cannot use "waking" as synonymous with "being awake." It is a quirk of English that methought to remedy by "wakeness."

[18] See also my discussion in *Neurophilosophy*.[11]

linked by micro techniques to interactions between specific circuits in the cortex and subcortical circuits, especially circuits in several key structures in the thalamus. Fifth, and more specifically, MEG data reveal an robust 40-Hz wave form during wakeness and dreaming.[36] The definition and amplitude is much attenuated during sleep, and the amplitude is modulated during wakeness and dreaming. Analysis of the wave by MEG reveals it to be a traveling wave, moving in the anterior to posterior direction in the brain, covering the distance in about 12 to 13 *milliseconds*. Cellular data suggest that these dynamical properties emerge from particular neural circuits and their dynamical properties.

What does all this add up to? Based on these data, and mindful of the various high-level data, Rodolfo Llinas and colleagues[36,37] have hypothesized that the fundamental organization subserving consciousness and the shifts seen in the SDW pattern are pairs of coupled oscillators, each of which connect thalamus and cortex, but each connects distinct cell populations via its own distinctive connectivity style (see Figure 2). One oscillator "family" connects neurons in a thalamic structure known as the intralaminar nuclei, a bagel-shaped structure whose neurons reach to the upper layers of cortex to provide a highly regular fanlike coverage of the entire cortical mantle. The other oscillator "family" connects neurons in thalamic nuclei for modality-specific information (MS nuclei) originating, for example, in the retina or the cochlea, with modality-specialized cortical areas (e.g., V2, S2). During deep sleep, the intralaminar neurons projecting to cortex cease their 40-Hz behavior. During deep sleep and dreaming, external signals to the cortex are gated by the reticular nucleus of the thalamus.

Ever so crudely, the idea is that the second oscillator "family" provides the content (visual, somatosensory etc.) while the first provides the integrating context. In deep sleep the oscillators are decoupled; in dreaming they are coupled but the MS oscillating circuit is largely nonresponsive to external signals from the periphery; in wakeness, the oscillators are coupled, and the MS circuit is responsive to external signals.

What are the effects of lesions to the intralaminar thalamic structures (bagel)? The main profile of small unilateral lesions is neglect (unawareness) of all stimuli originating the opposite body side.[52,54] Bilateral lesions result in "unarousability" meaning roughly that the patient initiates no behavior and responds very poorly to sensory stimuli or questions.[6,31] Animal studies show much the same profile.

Lesions to modality-specific regions of the thalamus, by contrast, lead to modality-specific losses in awareness—visual awareness, for example, will be lost, but awareness of sounds, touches etc. can be normal. Intriguingly, the MEGs of Alzheimer's patients who have degenerated to a state of inanition show a dilapidated 40-Hz wave form when it exists at all. Obviously these data are not decisive, but at least they are consistent with the hypothesis.

Do the Llinas hypothesis and the Crick hypothesis fit together? Minimally, they are consistent. Additionally, they are mutually supporting at the neuron and

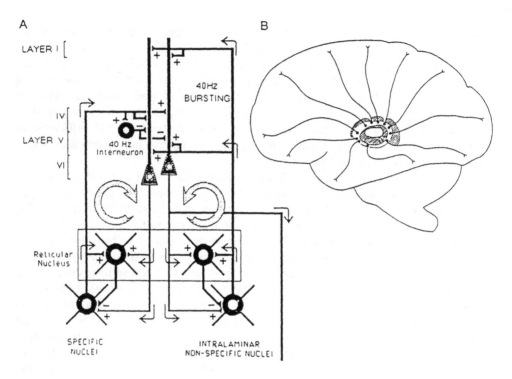

FIGURE 2 Schematic diagram of the circuits between the thalamus and the cerebral cortex proposed to serve temporal binding. (a) Diagram of two different types of circuit connecting thalamus and cortex. On the left, specific sensory nuclei or motor nuclei of the thalamus project to Layer IV of cortex, producing cortical oscillation by direct activation and feedforward inhibition via 40-Hz inhibitory interneurons. Collaterals of these projections produce thalamic feedback via the reticular nucleus (a kind of rind covering the thalamus). The return pathway (circular arrow with stipple) reenters this loop to specific and reticular nuclei via Layer VI cells. On the right, the second loop shows nonspecific intralaminar nuclei projecting to Layer I of cortex, and giving collaterals to the reticular nucleus. Layer V cells return oscillation to the reticular and intralaminar nuclei, establishing a second resonant loop. The conjunction of the specific and nonspecific loops is proposed to generate temporal binding. Connectivity between the loops is seen chiefly in Layer V. (b) Schematic diagram showing the intralaminar nuclei as a circular neuronal mass (stipple shading). Other parts of the thalamus are shown in hatched shading. The intralaminar nuclei project widely across the cortex, to Layer I. (From Llinas and Ribary.[37])

network level. One encouraging point is this: the two families of oscillators (MS and intralaminar) richly connect to each other mainly in *cortical layer 5* (see Figure 2). From what we can tell now, those connections seem to be the chief means whereby the oscillators are coupled. The possibility entertained here is that the temporal synchrony Crick hypothesizes in neurons carrying signals about external stimuli may be orchestrated by the intralaminar-cortical circuit. Connections between brainstem structures and the intralaminar nucleus could have a role in modulating arousal and alertness.

Many questions now suggest themselves. For example, how do the pivotal structures for awareness interface with behavior? More specifically, what are the connections between the intralaminar nuclei and motor structures, and between layer 5 of sensory corteces and motor structures; do the projections from the intralaminar nuclei to the cingulate cortex have a role in attention? These are questions motivated by independent data. Convergence of hypotheses is, of course, encouraging, but it is well to remember that it can also encourage us down the proverbial garden path. Wisdom counsels guarded optimism.

4. CONCLUDING REMARKS

Viewing matters from the mystery side of a phenomenon, solutions can seem impossible, and perhaps even unwanted. On the understanding side, however, solutions seem almost obvious and hard to miss. Why, one might wonder, did it take so long to figure out what the elements are? How could someone as brilliant as Aristotle miss the plausibility in Aristarchus' idea that the Earth was a sphere moving around the sun? The deeper truths are all too easy to miss, of course, just as it is all too easy for us to miss whatever it is that explains why animals sleep and dream, and what autism is. The problems for neuroscience and experimental psychology are hard, but, as we inch our way along and as new techniques increase noninvasive access to global brain processes in humans, intuitions change. What seems obvious to us was hot and surprising news only a generation earlier; what seems confounding to our imagination is routinely embraceable by the new cohort of graduate students. Who can tell with certainty whether or not all our questions about consciousness can eventually be answered? In the meantime, it is rewarding to see progress—to see some questions shift status from Mysteries We Can Only Contemplate in Awe, to Tough Problems We Are Beginning to Crack.

REFERENCES

1. Barnden, J. , S. Kankananhalli, and D. Dharmavaratha. "Winner-Take-All Networks: Time-Based Versus Activation-Based Mechanisms for Various Selection Tasks." In *Proceedings of the IEEE International Symposium on Circuits and Systems*, held in New Orleans, LA, 1990.
2. Berthier, N. E., S. P. Singh, A. G. Barto, and J. C. Houk. "A Cortico-Cerebellar Model that Learns to Generate Distributed Motor Commands to Control a Kinematic Arm." In *Neural Information Processing Systems 4*, edited by J. E. Moody, S. J. Hanson, and R. P. Lippmann, 611–618. San Mateo CA: Morgan-Kaufman, 1992.
3. Bickle, J. "Revisionary Physicalism." *Biol. & Phil.* **7** (1992): 411–430.
4. Block, Ned. "Consciusness Ignored? Review of D. C. Dennett's *Consciousness Explained*." *J. Phil.* **90(4)** (1993): 83–91.
5. Bogen, J. E. "Intralaminar Nuclei and the Where of Awareness." *Soc. Neurosci. Abs.* (1993).
6. Castaigne, P., F. Lhermitte, A. Buge, R. Escourolle, J. J. Hauw, and O. Lyon-caen. "Paramedian Thalamic and Midbrain Infarcts: Clinical and Neuropathological Study." *Ann. Neur.* **10** (1981): 127–148.
7. Churchland, P. M. *Matter and Consciousness*, 2nd ed. Cambridge, MA: MIT Press, 1988.
8. Churchland, P. M. "Betty Crocker's Theory of the Mind: A Review of *The Rediscovery of the Mind*, by John Searle." *London Review of Books* (1993).
9. Churchland, P. M. "Evaluating Our Self Conception." *Mind and Language* (1993).
10. Churchland, P. M., and P. S. Churchland. "Intertheoretic Reduction: A Neuroscientist's Field Guide." *Sem. Neurosci.* **4** (1990): 249–256.
11. Churchland, P. S. *Neurophilosophy*. Cambridge, MA: MIT Press, 1986.
12. Churchland, P. S. "Replies to Comments. Symposium on Patricia Smith Churchland's *Neurophilosophy*." *Inquiry* **29** (1987): 241–72.
13. Churchland, P. S. "Reduction and the Neurobiological Basis of Consciousness." In *Consciousness in Contemporary Science*, edited by A. J. Marcel and E. Bisiach, 273–304, 1988.
14. Churchland, P. S. "Can Neurogiology Teach Us Anything About Consciousness?" Presidential Address to the American Philosophical Association, Pacific Division. In *Proceedings and Addresses of the American Philosophical Association*, 23–40, Vol. 67(4). Lancaster, PA: Lancaster Press, 1994.
15. Churchland, P. S. *The Engine of the Sour.* Forthcoming.
16. Churchland, P. S., and T. J. Sejnowski. "Brain and Cognition." In *Foundations of Cognitive Science*, edited by M. Posner, 245–300. Cambridge, MA: MIT Press, 1989.
17. Churchland, P. S., and T. J. Sejnowski. *The Computational Brain*. Cambridge, MA: MIT Press, 1992.

18. Crick, F. H. C. *The Astonishing Hypothesis.* New York: Scribner's Sons, 1994.
19. Crick, F. H. C., and C. Koch. "Towards a Neurobiological Theory of Consciousness." *Sem. Neurosci.* **4** (1990): 263–276.
20. Damasio, A. R. *Descartes' Error.* New York: Putnam, 1994.
21. Damasio, H. "Neuroanatomy of Frontal Lobe *in vivo*: A Comment on Methodology." In *Frontal Lobe Function and Dysfunction*, edited by H. Levin, H. Eisenberg, and A. Benton, 92–101. New York: Oxford University Press, 1991.
22. Damasio, H., and A. R. Damasio. "The Neural Basis of Memory, Language and Behavioral Guidance: Advances with the Lesion Method in Humans." *Seminars in the Neurosciences* **4** (1990): 277–286.
23. Damasio, H., T. J. Grabowski et al. "Visual Recall With Eyes Closed and Covered Activated Early Visual Cortices." *Soc. Neurosci.Abs.* **(658.4)** (1993).
24. Dennett, D. C. *Consciousness Explained.* Boston: Little, Brown and Co., 1991.
25. Elman, J. L. "Distributed Representations, Simple Recurrent Networks, and Grammatical Structure." *Machine Learning* **7** (1991): 195–225.
26. Farah, M. J. "Neuropsychological Inference with an Interactive Brain: A Critique of the 'Locality Assumption.'" *Behav. & Brain Sci.* (1993).
27. Feyerabend, P. K. *Philosophical Papers*, Vols. 1 and 2. Cambridge: Cambridge University Press, 1981.
28. Flanagan, O. *Consciousness Reconsidered.* Cambridge, MA: MIT Press, 1992.
29. Flanagan, O. "Prospects for a Unified Theory of Consciousness, or, What Dreams are Made Of." In *Scientific Approaches to the Question of Consciousness: 25th Carnegie Symposium on Cognition*, edited by J. Cohen and J. Schooler. Hillsdale, NJ: L. Erlbaum.
30. Giles, C. L., C. B. Miller, D. Chen, G. Z. Sun, H. H. Chen, and Y. C. Lee. "Extracting and Learning and Unknown Grammar with Recurrent Neural Nets." In *Neural Information Processing Systems 4*, edited by J. E. Moody, S. J. Hanson, and R. P. Lippmann, 317–324. San Mateo CA: Morgan-Kaufman, 1992.
31. Guberman, A., and D. Stuss. "The Syndrome of Bilateral Paramedian Thalamic Infarction." *Neurology* **33** (1983): 540–546.
32. Henderson, V. W., M. P. Alexander, and M. A. Nalser. "Right Thalamic Injury, Impaired Visuospatial Perception and Alexia." *Neurology* **32** (1982): 235–240.
33. Jain, A. N. "Generalizing Performance in PARSEC—A Structured Connectionist Parsing Architecture." In *Neural Information Processing Systems 4*, edited by J. E. Moody, S. J. Hanson, and R. P. Lippmann, 209–216. San Mateo CA: Morgan-Kaufman, 1992.
34. Kosslyn, S. M. , N. M. Alpert, W. L. Thompson, V. Maljkovic, S. B. Weise, C. F. Chabris, S. E. Hamilton, S. L. Rauch, and F. S. Buoanno. "Visual Mental Imagery Activated Topographically Organized Visual Cortex: PET Investigations." *J. Cog. Neuroscience* **5** (1993): 263–287.

35. Lange, T. E. "Dynamically-Adaptive Winner-Take-All Networks." In *Neural Information Processing Systems 4*, edited by J. E. Moody, S. J. Hanson, and R. P. Lippmann, 341–348. San Mateo CA: Morgan-Kaufman, 1992.
36. Llinas R. R., and D. Pare. "Of Dreaming and Wakefulness." *Neuroscience* **44** (1991): 521–535.
37. Llinas, R. R., and U. Ribary. "Coherent 40-Hz Oscillation Characterizes Dream State in Humans." *Proc. Natl. Acad. Sci.* **90** (1993): 2078–2081.
38. Logothetis, N., and J. D. Schall. "Neural Correlates of Subjective Visual Perception." *Science* **245** (1989): 753-761.
39. Lycan W. G. *Consciousness.* Cambridge, MA: MIT Press, 1987.
40. McGinn, C. *The Problem of Consciousness.* Oxford: Blackwells, 1990
41. Mozer, M. C. "Induction of Multiscale Temporal Structure." In *Neural Information Processing Systems 4*, edited by J. E. Moody, S. J. Hanson, and R. P. Lippmann, 275–282. San Mateo CA: Morgan-Kaufman, 1992.
42. Mozer, M. C., and J. Bachrach. "SLUG: A Connectionist Architecture for Inferring the Structure of Fine-State Environments." *Machine Learning* **7** (1991): 139–160.
43. Pinkas, G. "Constructing Proofs in Symmetric Networks." In *Neural Information Processing Systems 4*, edited by J. E. Moody, S. J. Hanson, and R. P. Lippmann, 217–224. San Mateo CA: Morgan-Kaufman, 1992.
44. Pollack, J. B. "The Induction of Dynamical Recognizers." *Machine Learning* **7** (1991): 227–252.
45. Schaffner, K. F. "Theory Structure, Reduction, and Disciplinary Integration in Biology." *Biol. & Phil.* **8** (1993): 319–348.
46. Searle, J. R. *The Rediscovery of the Mind.* Cambridge, MA: MIT Press, 1992.
47. Singh, S. P. "The Efficient Learning of Multiple Task Sequences." In *Neural Information Processing Systems 4*, edited by J. E. Moody, S. J. Hanson, and R. P. Lippmann, 251–258. San Mateo CA: Morgan-Kaufman, 1992.
48. Sumida, R. A., and M. G. Dyer. "Propagation Filters in PDS Networks for Sequencing and Ambiguity Resolution." In *Neural Information Processing Systems 4*, edited by J. E. Moody, S. J. Hanson, and R. P. Lippmann, 233–240. San Mateo CA: Morgan-Kaufman, 1992.
49. Sutton, J. P., A. N. Mamelak, and J. A. Hobson. "Network Model of State-Dependent Sequencing." In *Neural Information Processing Systems 4*, edited by J. E. Moody, S. J. Hanson, and R. P. Lippmann, 283–290. San Mateo CA: Morgan-Kaufman, 1992.
50. Vergehsi, P., and D. G. Pelli. "The Information Capacity of Visual Attention." *Vision Research* **32** (1992): 983–995.
51. Viola, P. A. , S. G. Lisberger, and T. J. Sejnowski. "Recurrent Eye Tracking Network Using a Distributed Representation of Image Motion." In *Neural Information Processing Systems 4*, edited by J. E. Moody, S. J. Hanson, and R. P. Lippmann, 380–387. San Mateo CA: Morgan-Kaufman, 1992.
52. Watson, R. T., and K. M. Heilman. "Thalamic Neglect." *Neurology* **29** (1979): 690–694.

53. Watson, R. T., B. D. Miller, and K. M. Heilman. "Nonsensory Neglect." *Ann. Neurology* **3** (1978): 505–508.
54. Watson, R. T., E. Valenstein, and K. M. Heilman "Thalamic Neglect: Possible Role of the Medial Thalamus and Nucleus Reticularis in Behavior." *Archives of Neurology* **38** (1981): 501–506.

John F. Kihlstrom
Amnesia & Cognition Unit, Department of Psychology, University of Arizona, Tucson, AZ 85721
After August 15, 1994: Department of Psychology, Yale University, P. O. Box 208205, New Haven, CT 06520-8205

The Rediscovery of the Unconscious

To have ideas, and yet not be conscious of them,—there seems to be a contradiction in that; for how can we know that we have them, if we are not conscious of them? Nevertheless, we may become aware indirectly that we have an idea, although we be not directly cognizant of the same.

— Immanuel Kant[47]
Anthropology from a Pragmatic Point of View

Perhaps the greatest mystery of the human mind is consciousness: how it is that a physical system, composed of biological structures interacting according to chemical and electrical principles, gives rise to such subjective experiences as perceiving, remembering, thinking, feeling, wanting, and willing. But the mystery of consciousness does not rest solely on our incomplete knowledge of how bodily processes are related to mental states. It also reflects the fact that while we may have direct introspective access to our own minds, we can know the minds of other people only through their self-reports and behaviors—indices whose reliability is unknown in principle. Put another way: we have direct and irrefutable evidence of our own consciousness, but the attribution of consciousness to other people (and, for that matter, other animals) must remain an inference. Finally, even our knowledge of our own minds is likely to be incomplete, to the extent that our experiences, thoughts, and actions are governed by mental structures and processes that lie outside the scope of introspection.

These three problems—mind and body, other minds, and the unconscious mind—together summarize the scientific effort to understand the fundamental fact of human consciousness.[16,22,26,79] All three are important, but the one that interests me the most is the last one: whether it makes sense to talk about an unconscious mind and, if so, how best to characterize the relations between conscious and unconscious mental life.

1. PROLOGUE: THE DISCOVERY AND REDISCOVERY OF CONSCIOUSNESS

First, a little deep background. As everyone knows, scientific psychology began in the nineteenth century as the study of consciousness. The sensory psychophysics of Fechner[21] and Helmholtz,[30] and the experimental introspection of Wundt[94] and Titchener,[88] were fundamentally concerned with the analysis and determinants of conscious experience. William James, in *Principles of Psychology*,[43] wrote that "The first fact for us, then, as psychologists, is that thinking of some sort goes on" (p. 224); and in *Psychology: Briefer Course*,[44] James followed Ladd[62] in defining psychology as "the description and explanation of states of consciousness as such" (p. 1).

Soon, however, the stimulus-response connectionism of Thorndike[87] turned into the behaviorism of Watson,[90] which in turn quickly came to dominance, and psychology—so the joke goes—lost its mind. But not completely: interest in conscious experience was maintained by the Gestalt psychologists and others interested in visual perception; by the work of Woodworth[93] and others on the span of attention; and by the purposive psychology of McDougall.[68] What is now known as the cognitive revolution[3,24,39] changed all that: the revolution was ushered in by the work of Cherry[13] and Broadbent[8] on selective attention, which can be identified with consciousness; later, the multistore model of memory, popularized by Atkinson and Shiffrin,[2] essentially identified consciousness with primary or short-term memory, itself the product of selective attention. Kamin[46] showed that even something as elementary as classical conditioning in rats could not be understood without attributing mental states of surprise and expectancy to the animals. The rediscovery of consciousness was consolidated by the development of an experimental approach to mental imagery by Paivio[70] and Shepard,[81] among others.

This history is familiar to most people, but I review it here because it is important to understand that the cognitive revolution need not have revived interest in consciousness at all. Consider two other sources of the cognitive revolution: information-processing theory and modern linguistics. There is nothing about information processing per se that necessarily entails consciousness at any point in the sequence. Computers process information, and they have long served as models of the mind, but they are not conscious of the information that they process; in the most recent version of information-processing theory, variously known as connectionism (that Thorndikian term again!), parallel distributed processing, or neural network computation, consciousness is quite literally an afterthought.[67,76] Similarly, Chomsky[14] argued that language processing was performed by a set of structures and processes whose operation was completely inaccessible to consciousness, in principle. The listener does not consciously analyze the utterance of the speaker: all of this work is done automatically, by modules that have evolved for this specific purpose. Later, Fodor,[23] Jackendoff,[41] and others extended this notion of cognitively impenetrable modules to other domains, such as visual perception. The point is that cognition doesn't have to involve consciousness, and it is possible for respectable cognitive scientists to argue that consciousness has no functional significance at all—that it is a fixture of folk psychology that is better swept away.

2. THE INITIAL DISCOVERY OF THE UNCONSCIOUS

A somewhat similar account can be given of the discovery, loss, and rediscovery of *the psychological unconscious*—by which I mean not a place in the mind, but rather a domain of mental structures and processes which influence experience, thought, and action outside of phenomenal awareness and voluntary control.[50,52,53] The notion that unconscious processes are important elements of mental life is very old.[19,35,61,92] For example, in his *New Essays on Human Understanding*,[63] the German philosopher Leibnitz wrote about how our conscious thoughts are influenced by sensory stimuli of which we are not aware:

> ...at every moment there is in us an infinity of perceptions, unaccompanied by awareness or reflection.... That is why we are never indifferent, even when we appear to be most so.... The choice that we make arises from these insensible stimuli, which...make us find one direction of movement more comfortable than the other. (p. 53)

Furthermore, Kant's *Anthropology from a Pragmatic Point of View*,[47] his last work and perhaps the first comprehensive textbook of psychology, had a major section "Of the ideas which we have without being conscious of them" (pp. 18–20).

Inspired by Kant's distinction between *noumena* and *phenomena*, the unconscious was apparently a popular theme in nineteenth-century German philosophy, especially among the Romantic philosophers. Schopenhauer, in *The World as Will and Idea*,[78] argued that human thought and action was driven by unconscious, irrational instincts of conservation and sex. Herbart,[34] drawing on Leibnitz's views, described the sensory threshold, or *limen*, as a battleground where various perceptions competed for representation in consciousness. The stronger percepts pushed the weaker ones below the limen; but the repressed percepts continued to strive for expression, chiefly by associating themselves with other ideas. Even Marx and Engels get into the act: by means of mystification people hide, even from themselves, the true reasons for what they do; and by means of consciousness-raising, people become aware of the true nature of their current situation. This line of development reached its apex in von Hartmann's *Philosophy of the Unconscious.*[29] For Hartmann, the universe is ruled by the unconscious, a highly intelligent dynamic force composed of three layers: the absolute unconscious, accounting for the mechanics of the physical universe; the physiological unconscious, underlying the origin, development, and evolution of life; and the relative unconscious, which he considered to be the origin of conscious mental life.

> If we now institute a comparison between the Conscious and Unconscious, it is first of all obvious that there is a sphere which is always reserved to the Unconscious, because it remains for ever inaccessible to consciousness. Secondly, we find a sphere which in certain being only belongs to the Unconscious, but in others is also accessible to consciousness. Both the scale of organisms as well as the course of the world's history may teach us that all progress consists in magnifying and deepening the sphere open to consciousness; that therefore in *a certain* sense consciousness must be the higher of the two. Furthermore, if in man we consider the sphere belonging both to the Unconscious and also to consciousness, this much is certain, that everything which any consciousness has the power to accomplish can be executed equally well by the Unconscious, and that too always far more strikingly, and therewith far more quickly and more conveniently for the individual, since the conscious performance must be striven for, whereas the Unconscious comes of itself and without effort. (Hartmann,[29] Vol. 2, p. 39)

Hartmann's speculative philosophy was extremely popular—the three volumes, running to more than a thousand pages, went through a total of 12 editions. In the end it proved too speculative for the first generation of scientific psychologists—both Ebbinghaus[17] and James,[43] for example, roundly criticized it, not least because of the global, uncritical way in which it was applied:

> Hartmann fairly boxes the compass of the universe with the principle of unconscious thought. For him there is no nameable thing that does not exemplify it. But his logic is so lax and his failure to consider the most obvious alternatives so complete that it would, on the whole, be a waste of time to look at his arguments in any detail. (James,[43] p. 171)

Nevertheless, we owe to Hartmann the very concept of the psychological unconscious, as well as the Romantic notion, which is still with us today, that the unconscious is in some sense superior to consciousness. As Hartmann put it: "the Unconscious can really outdo all the performances of conscious reason" (Hartmann,[29]
Vol. 2, pp. 39–40).

Similarly, in the *Treatise on Physiological Optics*,[30] Helmholtz argued that
our conscious perceptions are determined by unconscious inferences (*umbewusster
Schluss*, literally "unconscious conclusion"), mental computations (as we would call
them today) of which we can never be aware, and over which we have no control:

> The psychic activities that lead us to infer that there in front of us at a
> certain place there is a certain object of a certain character, are generally
> not conscious activities, but unconscious ones. In their result they are the
> equivalent to *conclusion*, to the extent that the observed action on our
> senses enables us to form an idea as to the possible cause of this action.. ...
> But what seems to differentiate them from a conclusion, in the ordinary
> sense of that word, is that a conclusion is an act of conscious thought.. ...
> Still it may be permissible to speak of the psychic acts of ordinary percep-
> tion as *unconscious conclusions*.. ... (Helmholtz,[31] p. 174)

In later work, Helmholtz backed away from the label of unconscious conclusions,
because of its association with the Romantic unconscious of Schopenhauer and
Hartmann,[32] but he never abandoned the basic insight:

> I find even now that this name is admissible within certain limits since these
> associations of perceptions in the memory actually take place in such a man-
> ner, that at the time of their origin one is not aware of it.. ... (Helmholtz,[33]
> p. 255)

All of this laid the foundation for what Ellenberger[19] called *The Discovery
of the Unconscious* in psychiatry and psychology—a discovery that came in two
stages. What Ellenberger has called "the first dynamic psychiatry" covers the period
roughly between 1775 and 1900, beginning with the animal magnetism of Mesmer
and the hypnotism of Braid,[55] and culminating in the systematic study of hysteria
and multiple personality by Charcot[12] and Janet[45] in France, and Prince[74] and
Sidis[81] in America.

What Ellenberger[19] calls "the new dynamic psychiatry" was essentially the
creation of Freud, beginning with his collaborative *Studies on Hysteria*, published
with Breuer.[7] Based on their clinical observations, Breuer and Freud concluded
that the symptoms of hysteria were produced by unconscious memories of traumatic
events—"hysterics suffer mainly from reminiscences" (Breuer & Freud,[7] p. 7). These
events had been lost to conscious awareness, but nevertheless continued to influence
experience, thought, and action outside of awareness:

We may reverse the dictum *'cessante causa cessat effectus'* [when the cause ceases the effect ceases] and conclude from these observations that the determining process continues to operate in some way or other for years—not indirectly, through a chain of intermediate causal links, but as a *directly* releasing cause—just as a psychical pain that is remembered in waking consciousness still provokes a lachrymal secretion long after the event. *Hysterics suffer mainly from reminiscences.* (Breuer and Freud,[7] p. 7)

* * * * *

Our observations have shown...that the memories which have become the determinants of hysterical phenomena persist for a long time with astonishing freshness and with the whole of their affective colouring. We must, however, mention another remarkable fact...that these memories, unlike other memories of their past lives, are not at the patients' disposal. On the contrary, *these experiences are completely absent from the patients' memory when they are in a normal psychical state, or are only present in highly summary form.* Not until they have been questioned under hypnosis do these memories emerge with the undiminished vividness of a recent event. (Breuer and Freud,[7] p. 9)

Interestingly, in light of his critique of Hartmann,[29] James[43,86] himself was an active participant in the discovery of the unconscious. Although he disliked the notion of unconscious mental processes—for him, as for Searle,[79] consciousness and the mental were synonymous—James was persuaded by clinical observations of hysteria and hypnosis that even very complex mental processes could take place outside the scope of phenomenal awareness—a set of phenomena which he labeled *co-conscious* mental states.

Unfortunately, just when the concept of the psychological unconscious was getting up steam, the behaviorist revolution hit—and the psychological unconscious went the way of consciousness itself. It was bad enough to explain behavior in terms of mental states that could not be publicly observed; and so it was doubly bad to explain behavior in terms of mental states that could not even be *privately* observed! Interestingly, the rejection of unconscious mental processes was aided and abetted by James' critical remarks on Hartman's and other Romantic conceptions of the unconscious mind, as expressed in his view that the unconscious "is the sovereign means for believing what one likes in psychology, and of turning what might become a science into a tumbling-ground for whimsies" (James,[43] Vol. 1, p. 66).

Again, as with consciousness, some psychologists maintained an interest in the psychological unconscious. Unfortunately, by and large these individuals were psychoanalysts who isolated themselves from academic psychology, treating patients in their private offices and hospitals, training students in analytic institutes—a self-isolation that was reinforced by the prejudices of academic psychologists themselves.

There were exceptions: Rapaport[75] and others within the tradition of psychoanalytic ego psychology tried to maintain contact, and in the process conserved quite a bit of cognitive psychology against the behaviorists' hegemony; and the "New Look" of Bruner[9,10,11] and others tried to accomplish the same goal from the other side. Still, revival of academic interest in the psychological unconscious had to wait until the cognitive revolution was well consolidated. We are now at a point, however, where interest in the psychological unconscious runs wide and deep. This happy state of affairs is the end product of at least four quite independent strands of investigation, which together converge on our modern conception of the psychological unconscious.

3. AUTOMATIC AND STRATEGIC PROCESSING

One research tradition contributing to the modern interest in the psychological unconscious is the distinction commonly drawn between "automatic" and "strategic" cognitive processes. The concept is related to Helmholtz's notion of unconscious inference. Consider the *moon illusion*[48]: the moon on the horizon looks much larger than the moon at zenith. The explanation is that there is an inverse relationship between the retinal size of an object and its distance from the viewer. Because the background horizon appears to be farther away than the open sky, while the retinal image cast by the moon has not changed at all, the visual system calculates that the moon at horizon *must be* larger than the moon at zenith. This calculation is entirely unconscious, of course; and what is equally interesting, knowledge of the illusion does not diminish it at all. Another example of automaticity is found in the Stroop[85] *color-word effect*: a list of color names is printed in different colors: if the ink color matches the color name (e.g., the word *yellow* printed in yellow ink), naming the color of the ink is easy; but if the word and color do not match (e.g., *yellow* printed in green ink), it is very hard. Automatic decoding of the word interferes with naming of the color.

Some automatic processes seem to be innate, while others are automatized after extensive practice. Within broad limits,[4,64] both types of automatic processes appear to be inevitably engaged by the presentation of specific stimulus inputs; they are independent of any intentionality on the part of the subject; and they cannot be controlled or terminated before they have run their course. What interests me most, however, are the implications for consciousness. First, as Helmholtz noted, we have no conscious awareness of their operation. Second, we have little or no awareness of the information processed automatically.

Consider a study by Spelke, Hirst, and Neisser,[84] in which subjects were asked to read unfamiliar prose and take dictation at the same time. On initial trials, performance on both these attention-demanding tasks was seriously impaired: each required too much attention to be done at the same time as the other. After six weeks of practice, however, the subjects were able to take accurate dictation at the same time as they could read a prose passage (with 80% comprehension). Nevertheless, the subjects were generally unable to recall any of the words they had transcribed, and had little or no appreciation of how the lists had been structured. The dictation task, once automatized, no longer interfered with reading for comprehension; but neither did it yield memorable encodings of the dictated words.

In the world outside the laboratory, the flavor of this experience is captured beautifully in the following found poem, written by the American athlete and broadcast journalist Phil Rizzuto[72] (p. 7):

My Secret

When I'm driving
To Yankee Stadium and back,
I do it so often.

I don't remember passing lights.
I don't remember paying tolls
Coming over the bridge.

Going back over the bridge,
I remember...

Here we have someone, whose mind is focused elsewhere, performing a task he has accomplished hundreds and thousands of times in the past, apparently without any awareness of what he is doing.

While Hartmann thought that the progress of civilization consisted in expanding the range of consciousness, Alfred North Whitehead apparently argued the opposite[5]:

It is a profoundly erroneous truism...that we should cultivate the habit of thinking of what we are doing. The precise opposite is the case. Civilization advances by extending the number of important operations which we can perform without thinking about them. Operations of thought are like cavalry charges in a battle—they are strictly limited in number, they require fresh horses, and must only be made at decisive moments.

That quite complicated activities can be routinized, and performed without any concurrent awareness, is indicated by reports of three patients suffering from *petit mal* epilepsy:

One patient, whom I shall call A., was a serious student of piano...as apt to make a slight interruption in his practicing, which his mother recognized as the beginning of an "absence." Then he would continue to play for a time with considerable dexterity.... Sometimes the attack would come on [Patient B] while walking home from work. He would continue to walk and to thread his way through the busy streets on his way home. He might realize later that he had had an attack because there was a blank in his memory for a part of the journey. . . . If Patient C was driving a car, he would continue to drive, although he might discover later that he had driven through one or more red lights. (Pennfield,[71] p. 39)

Demonstrations of automaticity indicate that a great deal of complex cognitive activity can go on outside of conscious awareness—provided that the skills, rules, and strategies required by the task have been automatized. Previously, as in the Atkinson and Shiffrin[2] paradigm, unconscious preattentive processes were limited to elementary perceptual analyses of the physical features of environmental stimuli. Now it is clear that there are circumstances under which the meanings and implications of events can be unconsciously analyzed as well; moreover, adaptive behaviors can be organized in response to these events, all without these behaviors being represented in conscious awareness.

4. THE IMPACT OF COGNITIVE NEUROPSYCHOLOGY

About 20 years ago, as cognitive psychology turned into cognitive *neuro*psychology, researchers began to see evidence of the psychological unconscious in the behavior of brain-damaged patients. Pride of place, of course, goes to studies of the amnesic syndrome resulting from bilateral damage to the hippocampus and related structures in the medial temporal lobe, or, alternatively, the diencephalon and mammillary bodies. Such patients show a dense anterograde amnesia: after only a few moments of distraction, they cannot remember events that have just occurred. But as we all know, these patients also show the influence of the unremembered past on their current experience, thought, and action. For example, a patient who has recently seen the word *ASSASSIN* will be unable to recall or recognize the word shortly thereafter. But when asked to complete a fragment with a legal English word, they will be more successful with *A_A_I_* than with *T_P_R_*. This effect, known generically as priming, shows that something about the past event has been retained in memory, and actively influences current task performance.

Credit for this discovery goes to Warrington and Weiskrantz,[89] but the significance of their observation was not entirely clear until many years later. Based on effects such as these, Schacter[77] and others began to draw a distinction between two expressions of memory, explicit and implicit—or, alternatively, memory with and without awareness.[18,42] Explicit memory refers to one's conscious recollection of the past, as manifested on tasks like recall and recognition. Implicit memory refers to any change in experience, thought, and action that is attributable to a past event: priming effects, savings in relearning, and proactive and retroactive interference are good examples. We now know that explicit and implicit memory are dissociable in at least three different senses: (1) there are patients, such as amnesics, who show implicit memory in the absence of explicit memory; (2) there are some variables that influence explicit but not implicit memory, and others that influence implicit but not explicit memory, and others that influence both explicit and implicit memory, but in opposite directions; and (3) explicit and implicit memory are stochastically independent, in that priming effects occur for items that cannot be recognized.

Implicit memory can also be expressed psychophysiologically. Consider the syndrome of prosopagnosia, observed in patients who have suffered bilateral damage in the inferior portions of the . These individuals lose the ability to recognize faces of people who are objectively familiar to them. They can describe these faces accurately, and they can recognize the people in question from other physical features, such as voice or gait, but they can no longer put names to faces (including their own). But prosopagnosic patients show differential physiological responses to familiar and unfamiliar faces, and to correct vs. incorrect names paired with familiar faces; when asked whether two faces match, they do better with familiar than with unfamiliar ones; and concurrent presentation of familiar faces can influence the processing of words that are associated with these faces.[95] Again, all of these effects show that memory for the face, and the connection between facial and verbal knowledge, has been preserved at some level, even if the patients cannot express this knowledge consciously. By now, lots of evidence has accumulated in favor of the distinction between explicit and implicit memory.

An analogous distinction can be made in the domain of perception. Consider Weiskrantz's[91] classic studies of blindsight in patients with damage to the striate cortex of the occipital lobe. Such patients report a lack of visual experience in some portion of the visual field: when a stimulus is presented to their scotoma, they say they see nothing at all. Yet when forced to make guesses about the properties of a stimulus, their conjectures about presence, location, form, movement, velocity, orientation, and size prove to be more accurate than would be expected by chance alone. Something similar occurs in visual neglect arising from lesions in the temporoparietal region of one hemisphere (usually the right) that do not affect primary sensory or motor cortices. These patients appear to neglect that portion of the contralateral sensory field (usually the left). Thus, a patient asked to bisect a set of horizontal lines may ignore the ones on the left side of the page; and for the remainder, the pencil strokes tend to be located about one-quarter of the way in from the right. It is as if the left half of the page, and the left half of each line, is not seen at all. But it is, at least sometimes: thus Marshall and Halligan[66] showed a left-hemineglect patient two pictures of houses, one above the other. The pictures were identical in every respect, except that one had flames coming out of a window on the left side. The patient did not detect the difference, because after all the pictures had identical right sides; but when asked which she would prefer to live in, she consistently chose the one without the flames.

We may take studies such as these as evidence for implicit perception.[58] Whereas implicit memory refers to performance effects attributable to past events, implicit perception is concerned with the analogous effects of an event in the current environment, or the very immediate past. Of course, one could take all this just as evidence for implicit memory: after all, all introspection is retrospection, as someone

(William James, I think) once said. But there is an important difference between the two phenomena: in implicit memory, the events in question were consciously perceived at the time they occurred, and subsequently were lost to conscious recollection. In implicit perception the event was never consciously perceived at all: it is the perception that is implicit in task performance, not just the memory. In general, I prefer to reserve implicit memory for cases where the event in question was consciously perceived, but not consciously remembered.

Still, sometimes the distinction can get blurry. In an experiment performed with my colleagues Randall Cork and Daniel Schacter,[60] a group of patients received elective surgery with a mixture of isoflurane and oxygen as the anesthetic agent. During the operation, they heard a tape presenting a list of 15 paired-associates of the form *BREAD-BUTTER*. In the recovery room, and again 48 hours later, they were presented with the cue terms; on one test they were asked to produce the associated response, while on another they were asked to report the first word that came to mind. Cued recall was at chance levels, but free association showed a significant priming effect. Other people have gotten this kind of effect, too, but not everybody has been successful. And we ourselves failed to confirm the effect when we switched from isoflurane and oxygen to sufentanil and nitrous oxide as the anesthetic agent[15]—an outcome which if confirmed may begin to tell us something about the biological substrates of consciousness. But for the moment, assuming that the patients were unaware of the tape at the time it was presented, in this case we have implicit memory providing evidence for implicit perception.

---

5. SUBLIMINAL INFLUENCE

But you do not have to be brain-damaged to show perception and memory outside of conscious awareness. Consider a now-classic experiment by the philosopher-psychologist J. S. Pierce and his graduate student Joseph Jastrow[73]—Jastrow was the recipient of the first American Ph.D. in psychology,[6] and this study was probably the first psychological experiment performed in America. In a series of studies of weight and brightness discrimination, these experimenters reduced the difference between standard and comparison stimuli until they were at zero confidence in judging which was the heavier or brighter. Yet, when forced to guess, their choices were significantly more accurate than chance. Apparently, some stimulus information was registering somewhere in the perceptual system. Pierce and Jastrow thought their evidence disproved the very existence of a sensory threshold, or limen. That may be going too far, but more recent evidence confirms their essential point, that stimuli which cannot be consciously perceived can still have effects on conscious experience, thought, and action. Almost a century later, Marcel[65] revived the problem of subliminal perception—a special case of implicit perception[58] with his classic studies of priming in lexical decision. He found that prior presentation of words like DOCTOR primed lexical decisions—judgments of whether an item is a legal word—of words like NURSE, even though an intervening mask prevented conscious perception of the prime.

Early studies of subliminal perception, most often performed by psychoanalytic ego-psychologists or proponents of the New Look, were severely criticized on a variety of technical grounds, most of them now known to be misguided.[20,25,40] For example, it has been argued that any discriminative response is an indication of supraliminal stimulation, thus ruling out the notion of subliminal perception by fiat. Certainly one motive for the critique was the association of subliminal stimulation with the unconscious of psychoanalytic theory. But another reason was the simple fact that the theories of the time tended to describe cognition in terms of a series of ever more complicated processes, and thus had no room in them for the possibility that the meanings of words could be analyzed before the forms of words.

Things are different now. A great advance in this area was made by Merikle and his colleagues,[69] who distinguished between the *subjective threshold* (the point at which a stimulus cannot be consciously perceived) and the *objective threshold* (the point at which all differential response to a stimulus disappears). There is now considerable evidence from studies of identity priming (where, for example, TILE primes TILE), form priming (where FILE primes TILE), and semantic priming (where FILE primes INDEX) that subliminal perception is possible. This is especially true if presentation conditions are closer to the subjective threshold than to the objective threshold. But semantic priming is possible even under conditions that are near objective threshold. So, for example, Greenwald and his colleagues[28] found masked evaluative priming (e.g., where the connotative meaning of a word like JOY facilitates an evaluative judgment of a word like PEACE), under presentation conditions that were so degraded that the subjects were unable to guess *where* in the field the item was presented, much less its identity.

6. DISSOCIATIVE PHENOMENA IN HYPNOSIS

A final contribution in this area comes from research on hypnosis, a social interaction in which the subject acts on suggestions for experiences involving alterations in perception, memory, and the voluntary control of action. Many of these phenomena involve a division in consciousness, such that cognitive processing outside of phenomenal awareness influences ongoing experience, thought, and action.[35,50] In this case, however, the processes have not been routinized through repeated practice; and the percepts and memories in question are in no sense degraded.

Consider, for example, posthypnotic suggestion—the phenomenon that, so our mythology tells us, gave Freud his first good insight into the psychological unconscious. It may be suggested that, after the termination of hypnosis, the hypnotist will tap a pencil on the desk, at which time the subject will get up and sit in another, vacant chair; it may further be suggested that the subject will not remember receiving this suggestion, or anything else that has transpired while he or she was hypnotized. When the hypnotist taps a pencil, many subjects—especially those who are highly hypnotizable—will make the appropriate response, but deny knowing why they are doing so.[80] Even without the concomitant suggestion for amnesia, there is still something unconscious about the behavior: the subjects are aware of the significance of the cue, but they are not aware of any deliberate intention to move. Still, they move. When pressed, they will confabulate a reason, claiming that they wanted to stretch their legs, or that the upholstery was uncomfortable. This is a paradigm case of unconscious influence—the person does something without knowing why; but it is not an activity that has been practiced hundreds of times before; and there is nothing about the eliciting cue that is subliminal or otherwise degraded.

Posthypnotic response, especially when accompanied by posthypnotic amnesia,[51] involves a dissociation between explicit and implicit memory: the person's current behavior is influenced by a past event, in the absence of conscious awareness of that event. But other experiments illustrate the dissociation in more conventional terms. In one experiment,[49] hypnotized subjects memorized a list of words, and then received a suggestion for posthypnotic amnesia. After termination of hypnosis they received a recall test: Those subjects with the highest level of hypnotizability showed a very dense amnesia, compared to the performance of control subjects who were not deeply hypnotized. Later, all subjects were asked to give the first words that came to mind in response to various cues; some of these cues targeted the list items as free associates, others targeted control items, carefully matched to the list items, that had not been memorized. The nonamnesic subjects showed a substantial priming effect, producing more list than nonlist items as responses; but so did the amnesics, and in fact the magnitude of the priming effect was the same in the two groups. Interestingly, a second recall test continued to show a dense amnesia: apparently, production of list items as free associates did not remind the amnesic subjects of the items they had memorized. Finally, after the amnesia suggestion was canceled, everybody remembered the list almost perfectly.

This dissociation between explicit and implicit memory is quite different from the usual priming study, in a number of respects: good encoding was insured by requiring the subjects to meet a criterion of two perfect repetitions of the list before the amnesia suggestion was given; and adequate retention was demonstrated by the full recovery of memory after administration of the reversibility cue. Moreover, the priming observed here is semantic priming, not repetition priming: because the free-association cues were not presented as part of the study list, a semantic link between cue and target had to be formed by the subject at the time of encoding, and preserved in memory over the retention interval. Most explicit-implicit memory dissociations are a product of poor encoding; in this case, the reversibility of the amnesia indicates that it is a phenomenon of retrieval.

Hypnosis can give evidence of implicit perception, too, but again the details are different from the usual subliminal case.[58] Consider the familiar phenomenon of hypnotic analgesia,[36] in which the subject receives suggestions that he or she is insensitive to pain. The effect can be so profound as to permit highly hypnotizable patients to undergo major surgical procedures without benefit of chemical analgesia or anesthesia. But the analgesic patient's unawareness of pain does not mean that the pain has not registered in the perceptual-cognitive system. So, for example, physiological indices such as heart rate still respond to the pain stimulus, even though the subject reports feeling little or no pain.[37] Similarly, mental representations of the pain stimulus may be accessed, after analgesia has been successfully established, when the hypnotist attempts to communicate with a "hidden part" of the person that may know more than the "hypnotized part."[38] Under these circumstances, many analgesic subjects give pain reports that are comparable to those collected under normal, nonhypnotic conditions. Hilgard's[35] "hidden observer" is a metaphor for these mental representations of current and past experience, and the means by which they may be accessed.[54] The success of the hidden observer technique reflects implicit perception: despite their failure to experience pain, the pain stimulus has been registered and thoroughly processed by the sensory-perceptual system.

7. UNCONSCIOUS, PRECONSCIOUS, SUBCONSCIOUS

One of the major accomplishments of the cognitive revolution in psychology has been an increased appreciation of the role of unconscious processes in cognition, emotion, and motivation. It is now clear that even in the absence of conscious perception and memory, current and past events can influence the person's ongoing experience, thought, and action. The evidence for this conclusion provides the empirical basis for a provisional taxonomy of unconscious mental life.

First, there are a variety of cognitive processes which may be classified as *unconscious*, in the strict sense that they are inaccessible to phenomenal awareness under any circumstances, and can be known only by inference. The individual has no access to the rules by which these processes operate, or even any awareness that they are operating at all. Unconscious processes play a major role in mental life. They are the means by which we detect features and recognize patterns of stimulation,[2] and by which we execute cognitive and motoric skills.[1] According to the modal thinking in this area,[4,64,82] unconscious processes, aside from operating outside the scope of phenomenal awareness, are automatic and effortless—that is, they are inevitably invoked by particular stimulus inputs, and their execution consumes little or no attentional resources.

According to Helmholtz[30] and his progeny, unconscious processes are the stuff of which conscious experience is made. But implicit in this assertion is the idea that the declarative knowledge which these procedures generate is accessible to consciousness, even if the procedures themselves are not. But that turns out not to be the case. Conscious experience, thought, and action can be affected by representations, as well as processes, of which we are not aware. These mental contents might be called *preattentive*, but I prefer to honor Freud's legacy by labeling them *preconscious*. In contrast to unconscious mental processes, preconscious contents are available to phenomenal awareness, and accessible in principle, if not in fact. As a rule, preconscious processing—that is, the processing of preconscious declarative knowledge—reflects a degradation of stimulus input, trace information, or cognitive resources.

Preconscious processing often seems to be analytically limited. So, for example, the priming effects obtained in general anesthesia are of the repetition class, which can be mediated by a perceptual memory system that stores the results of structural, but not semantic, analyses. To take one example, it is difficult to show semantic (as opposed to repetition) priming effects in subliminal perception or amnesic patients. Such effects can be seen in other circumstances, however, such as hypnosis. The semantic priming observed in posthypnotic amnesia is quite different from what we observe in implicit memory associated with preconscious processing, because it seems to reflect the complex processing of events, including semantic analysis, and their retention over long periods of time; moreover, conscious access to these representations can be restored under appropriate conditions. We may follow Prince[74] in classifying these representations as subconscious, because they possess the attributes required to be fully conscious processes, but are nevertheless dissociated from the stream of phenomenal awareness.

What is the mechanism for this dissociation? Whatever it is, it does not appear to be quite the same as that involved in unconscious and preconscious processing. There is no extensive practice of a skill, leading to knowledge compilation or proceduralization. There is no degradation of perceptual representations, memory traces, or cognitive resources. The mental representations in question have been fully activated by perceptual inputs or acts of thought, well above any threshold that might be required for representation in working memory, and they are the products of complex analyses; yet they are still denied to introspective phenomenal awareness. Subconscious processing—that is, the processing of representations that should be conscious, but nevertheless are not—poses a difficult challenge for psychology.

My own view,[57] is that subconscious processes are mediated by multiple mental representations of the self,[59] each linked to a somewhat different set of experiences, thoughts, and actions, and that phenomenal awareness of percepts, memories, and thoughts depends on which self-representation resides in working memory at any particular time. Moreover, I think the connection to the self is also implicated in automatic and preconscious processing, so that it serves as a kind of final common pathway uniting all the various instances of unconscious mental life.

What really matters most, though, is that a century of scientific psychology, and a couple of decades of cognitive science, have moved us far beyond Hartmann's[29] speculative philosophy of the unconscious, and Freud's psychodynamic interpretations. We now know a great deal about the conditions under which mental processes may be automatized, and rendered unconscious. And we now know that we have to take such concepts as unconscious perception, memory, and thought seriously. We also know that Hartmann was wrong in his Romantic assertion that the Unconscious has the same power as conscious mental life. Everything we know about the psychological unconscious indicates that, for all its capacities, still it suffers from serious analytic limitations.[27,56]

The most important accomplishment, however, is that we now have good evidence, from a wide variety of research paradigms, that our experience, thought, and action is influenced by mental structures and processes of which we are not aware. The unconscious is not, as William James feared it would be, merely a "tumbling-ground for whimsies"; rather, it is an empirical fact of mind, and can be studied by the conventional techniques of psychological science. Thus, while we continue to work on the problem of mind and body, we must also strive to understand the nature of unconscious mental life and its relationship to consciousness.

ACKNOWLEDGMENTS

Paper presented at the Symposium on "The Mind, the Brain, and Complex Adaptive Systems," sponsored by the Krasnow Institute for Advanced Study of George Mason University, and the Santa Fe Institute, and held at George Mason University, Fairfax, Virginia, in May 1993. The point of view represented here is based on research supported by Grant No. MH-35856 from the National Institute of Mental Health. I thank John Allen, Terrence Barnhardt, Melissa Berren, Lawrence Couture, Elizabeth Glisky, Martha Glisky, Heather Law, Chad Marsolek, Shelagh Mulvaney, Victor Shames, Michael Valdiserri, and Susan Valdiserri for their comments.

This paper is dedicated to William E. Edmonston, Jr., on the occasion of his retirement from Colgate University.

Address correspondence to John F. Kihlstrom, Amnesia & Cognition Unit, Department of Psychology, University of Arizona, Tucson, Arizona 85721. E-mail: kihlstrom@ccit.arizona.edu or kihlstrom@arizvms.bitnet. After August 15, 1994: Department of Psychology, Yale University, P. O. Box 208205, New Haven, Connecticut 06520-8205.

REFERENCES

1. Anderson, J. R. "Acquisition of Cognitive Skill." *Psych. Rev.* **89** (1982): 369–406.
2. Atkinson, R. C., and R. M. Shiffrin. "Human Memory: A Proposed System and Its Control Processes." In *The Psychology of Learning and Motivation*, edited by K. W. Spence and J. T. Spence, Vol. 2, 89–195. New York: Academic Press, 1968.
3. Baars, B. J. *The Cognitive Revolution in Psychology.* New York: Guilford, 1986.
4. Bargh, J. A. "Conditional Automaticity: Varieties of Automatic Influence in Social Perception and Cognition." In *Unintended Thought*, edited by J. S. Uleman and J. A. Bargh, 3–51. New York: Guilford, 1989.
5. Barrow, J. D. *Pi in the Sky: Counting, Thinking, and Being.* New York: Oxford University Press, 1992.
6. Blumenthal, A. L. "The Intrepid Joseph Jastrow." In *Portraits of Pioneers in Psychology*, edited by G. Kimble, M. Wertheimer, and C. White, 75–87. Hillsdale, NJ: Erlbaum, 1991.
7. Breuer, J., and S. Freud. "Studies on Hysteria." In *The Standard Edition of the Complete Psychological Works of Sigmund Freud*, edited by J. Strachey, Vol. 2. London: Hogarth Press, 1955.
8. Broadbent, D. E. *Perception and Communication.* London: Pergamon Press, 1958.
9. Bruner, J. "Another Look at New Look 1." *Am. Psychol.* **47** (1992): 780–783.
10. Bruner, J. "The View from the Heart's Eye." In *The Heart's Eye: Emotional Influences in Perception and Attention*, edited by P. M. Niedenthal and S. Kitayama, 269–286. San Diego, CA: Academic Press, 1994.
11. Bruner, J. S., and L. Postman. "Emotional Selectivity in Perception and Reaction." *J. Personality* **16** (1947): 69–77.
12. Charcot, J.-M. *Clinical Lectures on Diseases of the Nervous System.* London: Tavistock/Routledge, 1991. Originally published, 1877.
13. Cherry, E. C. "Some Experiments on the Recognition of Speech, with One and with Two Ears." *J. Acoust. Soc. Am.* **25** (1953): 975–979.
14. Chomsky, N. "Language and Unconscious Knowledge." In *Rules and Representations*, 217–254. New York: Columbia University Press, 1980.
15. Cork, R. C., J. F. Kihlstrom, and D. L. Schacter. "Absence of Explicit and Implicit Memory with Sufentanil/Nitrous Oxide." *Anesthesiology* **76** (1992): 892–898.
16. Dennett, D. *Consciousness Explained.* Boston: Little, Brown, 1991.
17. Ebbinghaus, H. "On Hartmann's Philosophy of the Unconscious." Ph.D. thesis, University of Bonn, 1873.
18. Eich, E. "Memory for Unattended Events: Remembering With and Without Awareness." *Memory & Cognition* **12** (1984): 105–111.

19. Ellenberger, H. *The Discovery of the Unconscious: The History and Evolution of Dynamic Psychiatry.* New York: Basic Books, 1970.
20. Eriksen, C. W. "Discrimination and Learning Without Awareness: A Methodological Survey and Evaluation." *Psych. Rev.* **67** (1960): 279–300.
21. Fechner, G. T. *Elements of Psychophysics*, Vol. 1. New York: Holt, Rinehart, and Winston, 1966. Original work published 1860.
22. Flanagan, O. *Consciousness Reconsidered.* Cambridge, MA: MIT Press, 1992.
23. Fodor, J. A. *The Modularity of Mind.* Cambridge, MA: MIT Press, 1983.
24. Gardner, H. *The Mind's New Science: A History of the Cognitive Revolution.* New York: Basic Books, 1985.
25. Goldiamond, I. "Indicators of Perception: 1. Subliminal Perception, Subception, Unconscious Perception: An Analysis in Terms of Psychophysical Indicator Methodology." *Psychol. Bull.* **55** (1958): 373–411.
26. Goldman, A. I. "Consciousness, Folk Psychology, and Cognitive Science." *Consciousness & Cognition* **2** (1993): 364–382.
27. Greenwald, A. G. "New Look 3: Unconscious Cognition Reclaimed." *Am. Psychol.* **47** (1992): 766–779.
28. Greenwald, A. G., M. R. Klinger, and T. J. Liu. "Unconscious Processing of Dichotically Masked Words." *Memory & Cognition* **17** (1989): 35–47.
29. Hartmann, E. V. *Philosophy of the Unconscious: Speculative Results According to the Inductive Method of Physical Science.* London: Routledge and Kegan Paul, 1931. Original work published in 1868.
30. Helmholtz, H. V. *Treatise on Physiological Optics*, 3 vols. Rochester, NY: Optical Society of America, 1924. Originally published 1866.
31. Helmholtz, H. V. "Concerning the Perceptions in General." In *Helmholtz on Perception, Its Physiology, and Development*, edited by R. M. Warren and R. P. Warren, 171–203. New York: Wiley, 1968. Original work published 1866.
32. Helmholtz, H. V. "The Facts of Perception." In *Helmholtz on Perception, Its Physiology, and Development*, edited by R. M. Warren and R. P. Warren, 249–260. New York: Wiley, 1968. Original work published 1878.
33. Helmholtz, H. V. "The Origin of the Correct Interpretation of Our Sensory Impressions." In *Helmholtz on Perception, Its Physiology, and Development*, edited by R. M. Warren and R. P. Warren, 249–260. New York: Wiley, 1968. Original work published 1894.
34. Herbart, J. F. *A Textbook in Psychology: An Attempt to Found the Science of Psychology on Experience, Metaphysics, and Mathematics.* New York: Appleton, 1891. Original work published 1816.
35. Hilgard, E. R. *Divided Consciousness: Multiple Controls in Human Thought and Action.* New York: Wiley-Interscience, 1977.
36. Hilgard, E. R., and J. R. Hilgard. *Hypnosis in the Relief of Pain.* Los Altos, CA: Kaufman, 1975.
37. Hilgard, E. R., A. H. Morgan, A. F. Lange, J. R. Lenox, H. Macdonald, G. D. Marshall, and L. B. Sachs. "Heart Rate Changes in Pain and Hypnosis." *Psychophysiology* **11** (1974): 692–702.

38. Hilgard, E. R., A. H. Morgan, and H. Macdonald. "Pain and Dissociation in the Cold Pressor Test: A Study of Hypnotic Analgesia with 'Hidden Reports' Through Automatic Key Pressing and Automatic Talking." *J. Abnormal Psychol.* **84** (1975): 280–289.

39. Hirst, W., ed. *The Making of Cognitive Science: Essays in Honor of George A. Miller.* Cambridge: Cambridge University Press, 1988.

40. Holender, D. "Semantic Activation Without Conscious Identification in Dichotic Listening, Parafoveal Vision, and Visual Masking: A Survey and Appraisal." *Behav. & Brain Sci.* **9** (1986): 1–23.

41. Jackendoff, R. *Consciousness and the Computational Mind.* Cambridge, MA: MIT Press, 1987.

42. Jacoby, L. L., and M. Dallas. "On the Relationship Between Autobiographical Memory and Perceptual Learning." *J. Exp. Psych.: General* **110** (1981): 306–340.

43. James, W. *Principles of Psychology.* Cambridge, MA: Harvard University Press, 1981. Original work published 1890.

44. James, W. *Psychology: Briefer Course.* Cambridge, MA: Harvard University Press, 1981. Original work published 1892.

45. Janet, P. *The Major Symptoms of Hysteria.* New York: Macmillan, 1907.

46. Kamin, L. J. "Predictability, Surprising, Attention, and Conditioning." In *Punishment and Aversive Behavior*, edited by B. A. Campbell and R. M. Church, 279–296. New York: Appleton-Century-Crofts, 1969.

47. Kant, I. *Anthropology from a Pragmatic Point of View.* Carbondale, IL: Southern Illinois University Press, 1977. Original work published 1798.

48. Kaufman, L., and I. Rock. "The Moon Illusion." *Science* **136** (1962): 953–961.

49. Kihlstrom, J. F. "Posthypnotic Amnesia for Recently Learned Material: Interactions with 'Episodic' and 'Semantic' Memory." *Cog. Psychol.* **12** (1980): 227–251.

50. Kihlstrom, J. F. "Conscious, Subconscious, Unconscious: A Cognitive Perspective." In *The Unconscious Reconsidered*, edited by K. S. Bowers and D. Meichenbaum, 149–211. New York: John Wiley and Sons, 1984.

51. Kihlstrom, J. F. "Posthypnotic Amnesia and the Dissociation of Memory." In *The Psychology of Learning and Motivation*, edited by G. H. Bower, Vol. 19, 131–178. New York: Academic Press, 1985.

52. Kihlstrom, J. F. "The Cognitive Unconscious." *Science* **237** (1987): 1445–1452.

53. Kihlstrom, J. F. "The Psychological Unconscious." In *Handbook of Personality: Theory and Research*, edited by L. Pervin, 445–464. New York: Guilford, 1990.

54. Kihlstrom, J. F. "Dissociation and Dissociations: A Comment on Consciousness and Cognition." *Consciousness & Cognition* **1** (1992a): 47–53.

55. Kihlstrom, J. F. "Hypnosis: A Sesquicentennial Essay." *Intl. J. Clinical & Exper. Hypnosis* **40** (1992b): 301–314.

56. Kihlstrom, J. F. "The Continuum of Consciousness." *Consciousness & Cognition* **2** (1993): 334–354.
57. Kihlstrom, J. F. "Consciousness and Me-ness." In *Scientific Approaches to the Question of Consciousness*, edited by J. Cohen and J. Schooler. Hillsdale, NJ: Erlbaum, 1994 (in press).
58. Kihlstrom, J. F., T. M. Barnhardt, and D. J. Tataryn. "Implicit Perception." In *Perception Without Awareness*, edited by R. F. Bornstein and T. S. Pittman, 17–54. New York: Guilford, 1992.
59. Kihlstrom, J. F., and S. B. Klein. "The Self as a Knowledge Structure." In *Handbook of Social Cognition*, edited by R. S. Wyer and T. K. Srull, 2nd ed., Vol. 1, 153–208. Hillsdale, NJ: Erlbaum, 1994.
60. Kihlstrom, J. F., D. L. Schacter, R. C. Cork, C. A. Hurt, and S. E. Behr. "Implicit and Explicit Memory Following Surgical Anesthesia." *Psychol. Sci.* **1** (1990): 303–306.
61. Klein, D. B. *The Unconscious: Invention or Discovery? A Critico-Historical Inquiry.* Santa Monica, CA: Goodyear, 1977.
62. Ladd, G. T. *Outlines of Physiological Psychology.* New York: Scribner's, 1891.
63. Leibniz, G. W. *New Essays on Human Understanding.* Cambridge: Cambridge University Press, 1981. Original work published 1704.
64. Logan, G. D. "Automaticity and Cognitive Control." In *Unintended Thought*, edited by J. S. Uleman and J. A. Bargh, 52–74. New York: Guilford, 1989.
65. Marcel, A. "Conscious and Unconscious Perception: Experiments on Visual Masking and Word Recognition." *Cog. Psychol.* **15** (1983): 197–237.
66. Marshall, J. C., and P. W. Halligan. "Blindsight and Insight in Visuo-Spatial Neglect." *Nature* **336** (1988): 766–767.
67. McClelland, J. L., D. E. Rumelhart, and the PDP Research Group. *Parallel Distributed Processing: Explorations in the Microstructures of Cognition, Vol. 2: Psychological and Biological Models.* Cambridge, MA: MIT Press, 1986.
68. McDougall, W. *Outline of Psychology.* New York: Scribners, 1923.
69. Merikle, P. M., and E. M. Reingold. "Measuring Unconscious Perceptual Processes." In *Perception Without Awareness*, edited by R. F. Bornstein and T. S. Pittman, 55–80. New York: Guilford, 1992.
70. Paivio, A. *Imagery and Verbal Processes.* New York: Holt, Rinehart, and Winston, 1971.
71. Penfield, W. *The Mystery of the Mind: A Critical Study of Consciousness and the Human Brain.* Princeton: Princeton University Press, 1975.
72. Peyer, T., and H. Seely, eds. *O Holy Cow! The Selected Verse of Phil Rizzuto.* Hopewell, NJ: Ecco Press, 1993.
73. Pierce, C. S., and J. Jastrow. "On Small Differences in Sensation." *Mem. NAS* **3** (1884): 75–83.
74. Prince, M. *The Unconscious: The Fundamentals of Human Personality Normal and Abnormal.* New York: Macmillan, 1914.
75. Rapaport, D., ed. *Organization and Pathology of Thought: Selected Sources.* New York: Columbia University Press, 1960.

76. Rumelhart, D. E., J. L. McClelland, and the PDP Research Group. *Parallel Distributed Processing: Explorations in the Microstructures of Cognition, Vol. 1: Foundations.* Cambridge, MA: MIT Press, 1986.

77. Schacter, D. L. "Implicit Memory: History and Current Status." *J. Exp. Psych.: Learning, Memory, and Cognition* **13** (1987): 501–518.

78. Schopenhauer, A. *The World as Will and Idea.* London: Routledge and Kegan Paul, 1964. Original work published 1819.

79. Searle, J. *The Rediscovery of Mind.* Cambridge, MA: MIT Press, 1992.

80. Sheehan, P. W., and M. T. Orne. "Some Comments on the Nature of Posthypnotic Behavior." *J. Nervous & Mental Disease* **146** (1968): 209–220.

81. Shepard, R. N. "The Mental Image." *Am. Psychol.* **33** (1978): 125–137.

82. Shiffrin, R. M. "Attention." In *Stevens' Handbook of Experimental Psychology,* edited by R. C. Atkinson, R. J. Herrnstein, G. Lindzey, and R. D. Luce, 2nd ed., Vol. 2, 739–812. New York: Wiley-Interscience, 1988

83. Sidis, B. *Psychopathological Researches: Studies in Mental Dissociation.* New York: Stechert, 1902.

84. Spelke, E., W. Hirst, and U. Neisser. "Skills of Divided Attention." *Cognition* **4** (1976): 215–230.

85. Stroop, J. R. "Studies of Interference in Serial Verbal Reactions." *J. Exp. Psych.* **18** (1935): 643–662.

86. Taylor, E. *William James on Exceptional Mental States: The 1896 Lowell Lectures.* New York: Scribners, 1982.

87. Thorndike, E. L. *Educational Psychology.* New York: Teachers College, 1913–1914.

88. Titchener, E. B. *A Textbook of Psychology.* New York: Macmillan, 1910.

89. Warrington, E. K., and L. Weiskrantz. "New Method of Testing Long-Term Retention with Special Reference to Amnesic Patients." *Nature* **217** (1968): 972–974.

90. Watson, J. B. "Psychology as the Behaviorist Views It." *Psych. Rev.* **20** (1913): 158–177.

91. Weiskrantz, L. *Blindsight: A Case Study and Implications.* Oxford: Oxford University Press, 1986.

92. Whyte, L. L. *The Unconscious Before Freud.* New York: Basic Books, 1960.

93. Woodworth, R. S. *Experimental Psychology.* New York: Holt, 1938.

94. Wundt, W. *Principles of Physiological Psychology.* New York: Macmillan, 1904. Original work published 1873–1874.

95. Young, A. W., and E. H. F. de Haan. "Face Recognition and Awareness After Brain Injury." In *The Neuropsychology of Consciousness,* edited by A. D. Milner and M. D. Rugg, 69–90. London: Academic Press, 1992.

David E. Rumelhart
Department of Psychology, Stanford University, Stanford, CA 94305

Affect and Neuro-Modulation:
A Connectionist Approach

1. INTRODUCTION

Perhaps the key question in psychology and neuroscience today is the relationship between the mind and the brain.[1] On the face of it, the mind and brain have almost nothing in common. The mind seems to be ephemeral and abstract, whereas the brain is physical and mechanical. Nevertheless, it is the strong hypothesis of modern psychology, philosophy, and biology that the mind is a reflection of the behavior of the brain. It is all the more puzzling because we seem to have some independent access of the workings of our minds, but the behavior of the brain can only be studied through external means.

At various points during the history of psychology, it has been supposed that the mind could be studied and understood independently of the brain. It has sometimes been supposed that there is a psychological reality that can be wholly studied and understood in its own realm and on its own terms without consideration of the brain. On this account, the mind, roughly, is the "software" and the brain the "hardware."

---

[1]The ideas presented in this paper are the result of a series of discussions with a number of graduate students at Stanford University. The group included Brian Knutson, Monisha Pasupathi, Michael Fleming, Kenneth Kurtz, Elizabeth Olds, and Christopher Dryer

We can, it is assumed, understand the nature of the mind by understanding the "program" the brain is running. The details of how the brain works is, on this view, not important to the understanding of the important mental tasks humans carry out. Although it is nice to understand the workings of the brain, it is not necessary for a full understanding of how people think, reason, and engage in mental life.

In contrast to this, there is good evidence that the mind cannot be clearly understood without consideration of how the brain works. Perhaps the clearest evidence comes from cases of brain damage in which the mind breaks up in often surprising ways. The most natural tasks become impossible, personalities change, people cannot talk, or cannot remember or cannot reason clearly. There are numerous cases in the clinical literature of people whose deficits are incredibly specific amd difficult to explain. In addition to changes due to physical damage, there are many examples of "reversible" changes resulting from drugs of various kinds. In a way, these cases make it most clear that the mind is a product of brain. Drugs alone can do most of the things that direct damage does to the brain. They can change personalities, effect our ability to remember and reason, make us angry or happy, etc. etc. In short, virtually anything that effects our brain also effects our mind—often in ways that would seem to be difficult to predict. The deepest aspects of who we are can, it would seem, just change as our brain is impinged upon.

The connectionist, PDP, or neural network approach to the study of mind and brain constitutes one possible means of attacking this problem. In this case, we attempt to model the relationship between mind and brain through the development of computational models which, on the one hand, attempt to model important aspects of the way in which brains work, while at the same time behaving in ways consistent with human behavior. There is a large literature on applications of this kind. In the sections below, I will first provide a standard connectionist model of the relationship between mind and brain and then, provide an elaboration of that model to attempt to explain certain affective phenomena.

2. THE CANONICAL CONNECTIONIST MODEL

In Figure 1 we show the canonical connectionist model. The elements of the model is to be understood in the following way:

1. The Units (dots in the figure) represent "hypotheses" about the world.
2. Each unit has an Activation level which represents the "confidence" that the hypothesis is true.
3. The connections among units, represented by weights (e.g., W_{ij}) correspond to "constraints" among the hypotheses. For each pair of units there is a weight which represents the constraints between the units. The weights can be positive or negative, large or small. A large positive weight between two units represents a large constraint between the two units. That is, it represents the idea that

when the hypothesis corresponding to one of the units is true, then it is likely that the hypothesis corresponding to the other unit is also true. A large negative weight indicates that when one hypothesis is true, it is likely that the other is false. Smaller weights correspond to smaller constraints.

4. The Inputs constitute the external evidence to the network.
5. The Outputs provide the "proposed action" of the network.
6. The Final State to which the network settles represents the "interpretation" of the external events.
7. Each unit has a bias term, $(\beta|i)$ which represents the "prior" or default probability that the hypothesis is true.

The network works in the following way. Some input signal imposes a pattern of activity over the set of input units. These units are connected to other units and cause activity to arise in these other units. As each unit impinges on its neighbor, the pattern of activity is modified depending on the pattern of weights in the network. Since every pair of units are reciprocally connected, eventually the network reaches an equilibrium state and stops changing. At this point the network can be said to have "settled" into an "interpretation" of the input pattern.

Let the activation of a given unit be

$$a_i = \frac{1}{1 + e^{\eta_i}} \tag{1}$$

where $\eta_i = \sum_j w_{ij}a_j + \beta_i$. We then can get a measure of the degree to which the overall network has achieved a "good" interpretation of the input by computing what is called the "harmony" of the network, $\mathcal{H}$, where

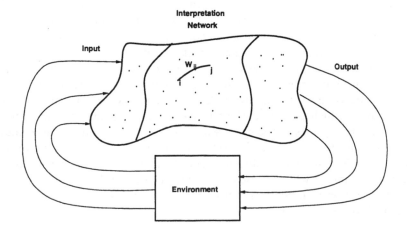

FIGURE 1   The canonical connectionist model.

$$\mathcal{H} = \sum_{ij} w_{ij} a_i a_j + \sum_i a_i \beta_i \tag{2}$$

In general, the network moves in such a way as to maximize harmony.

3. CONNECTIONIST MODELS AND NEUROMODULATORS

In a typical connectionist network, there is a single weight connecting any two units. In the brain, however, we know that there are many different kinds of connections and, under certain conditions, a particular unit may have more than one neurotransmitter connecting a pair of units. One simple example is illustrated below. This represents a case in which there are multiple neurotransmitters in a single neuron. In this case, depending on the hormone level of a female rat, it was determined that there were one of two different neurotransmitters (or neuropeptides) generated. It is useful to see the consequences of such a system. In one hormonal state, say, a high hormonal state, the system generates mostly "type x" transmitter. In the second, low hormonal state, the system generates mostly "type o" transmitter. Now, suppose that different receiving synapses have different kinds of receptors. In that case, when hormone levels are high, those synapses that are sensitive to "type x" neurotransmitter will be active while those sensitive to "type o" transmitters will be quiet. Learning will occur at the "active" receptor sites. On the other hand, when hormone levels are low, the animal is in "state 2," and learning will occur at the other synapse. Now since, in connectionist models, learning is dependent on the activity at the synapse, we expect that different behaviors will be learned in the different states. In this case, that would amount to the animal learning different behaviors depending on the hormone level, a kind of state-dependent learning. It could, it would seem, be a reasonable mechanism for the animal to learn different behaviors under different hormone-dependent conditions.

In the model to be described below, we generalize this idea in the following way. We imagine that for each unit, there is, in general, a number of possible types of neuropeptides or neurotransmitters. Certain neurons are assumed to generate neuromodulators which determine the relative amounts of the different neurotransmitters and neuropeptides in the synapses. A particular neuromodulator, indicated by α_k, is assumed to determine the relative amounts of a particular transmitter. Moreover, there is assumed another neuromodulator, the "gain" term (c.f. Cohen and Servin-Schreiber[3]) which is associated with the general level of arousal in the person. These terms combine to determine the activity level of a particular unit. The activity level is given by

$$a_i = \frac{1}{1 + e^{\eta(i)}} \tag{3}$$

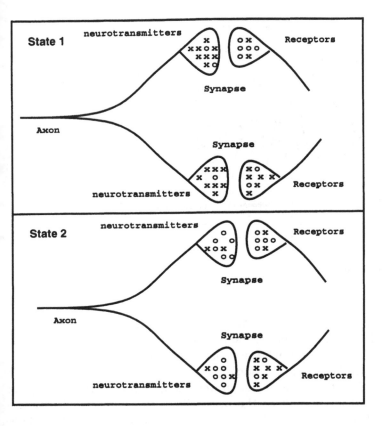

FIGURE 2
Multistate
units.

where $\eta_i = g \sum_j (\sum_k w_{ijk} \alpha_k) + \beta_i$ and g is a multiplicative "gain" term which is assumed to be proportional to the general arousal level in the system.

4. AFFECT AND STATE-DEPENDENT LEARNING

Figure 3 illustrates a more general learning system. In this case, we imagine that there a set of special units (indicated in the boxes) which produce neuromodulators which in turn determine the strength of each of the different α_i in the system. We can imagine that the system works in the following way. When certain special stimuli activate the system, they cause an increase in the amount of neuromodulator generated.

Imagine, as a simple case, that a child is being held in its mother's arms and that this generates a kind of "primitive" pleasure response in the child. This response, we imagine, directly stimulates certain aspects of the "affective" system

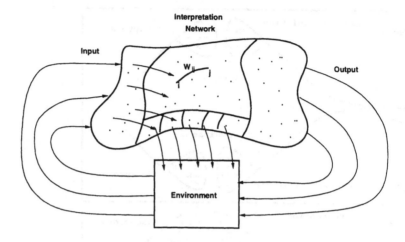

FIGURE 3   Multistate units.

illustrated by the arrows in the boxes. As they stimulate this system, there is a global increase in the amount of the relevant neuromodulator in the system. This leads to a strengthening of those weights associated with the pleasure response and, perhaps, a weakening of those weights associated with other responses. As always, in our connectionist models, there is a change in weights as a result of the experience; however, the weights that are changed the most will be those associated with the particular neuromodulators relevant to these pleasure responses. In general, these changes take place throughout the network and so there is a particular learning event that has taken place relevant to this particular situation. Since there are symmetric connections to and from the neuromodulatory system, activity in the rest of the network can cause the modulatory system to become active and, in that way, initiate the relevant affective state. Thus, in the future, it can happen that events which occurred in the context of the "primitive" pleasure stimulation can themselves initiate a kind of positive response. Of course, the same thing can happen with negative responses.

5. SET POINTS AND EQUILIBRATION

We imagine that for each neuromodulator type, there is a particular default quantity which constitutes the neutral point for the system. Stimulation of one sort or another moves the system away from the neutral point. Figure 4 illustrates this general idea. For each neuromodulator (α_i), there is a default amount that is

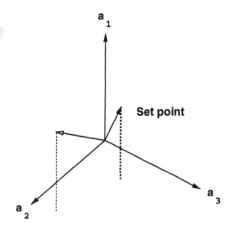

FIGURE 4　Set point and equilibrium.

generated by the system. The default amount constitutes the basic "set point" for the system, however, since each experience we have may activate certain of the neuromodulatory units and, thereby, move us away from the global set point. The white arrow is to represent an affective state at some point in time. We assume that there is a mechanism that moves us back to the equilibrium condition. That is, after some time, the white arrow moves toward the black arrow labeled set point. Each of the neuromodulatory units are assumed to be sensitive to the particular neuromodulator generated by the unit and, if there is too much, the unit reduces the amount it generates. If there is too little, it increases it. It is assumed to do this by varying the β term. Let $\dot{\alpha}_i$ be the target value of for neuromodulator α_i. Then, we suppose that the bias β_i is changed according to the following rule: $\delta\beta_i = \gamma(\dot{\alpha}_i - \alpha_i)$ where γ determines the rate of change. In this way, should the input remain constant for a period of time, the network will eventually converge at the neutral point. There are a number of interesting consequences of this phenomenon. In particular, it suggests that affective states are all relative. That is, up to a point, if things are generally positive the system will acclimate to the positive state and and a positive affect will occur only if the situation is better than average. Similarly, with negative states, the key is whether things are better or worse than average.

Another reasonably obvious consequence of this system comes when an external source provides the input. Consider, for example, the role of endorphins. Normally, when we receive a painful stimulation the system increases the quantities of endorphins in the system and, in that way, reduces the pain. If the pain continues, endorphins are produced at such a level as to essentially cancel out the pain (up to a point). Now imagine that some external source of an endorphin-like substance, such as morphine, is externally available. In this case, as the amount of the morphine increases, the units will detect the endorphin-like substance and stop generating endorphins. Then, when the morphine is removed, it can take the system some time to generate the endorphins. In the mean time, there may be substantial discomfort. Morphine addiction may well result from a mechanism such as this.

6. GLOBAL AFFECT

One interesting question about the role of affect is the degree to which it is a global phenomenon. On our account, it is a global phenomenon. We should find that when in a negative mood state, it should be more likely than not that other things will be viewed negatively. The reason for this is that when in a negative state, the weights associated with negative experiences will be the strongest and, when in a positive state, the weights associated with positive experiences will be the strongest. There are a number of sources of evidence consistent with the expectation. For example, when in a happy mood, people are more likely to see things as positive and over estimate the liklihood of positive events and underestimate the probability of negative events. Exactly the opposite occurs when in a negative state (c.f. Bower[1]). Similarly, when in a happy mood, people are more likely to remember happy memories from their childhood and, when in a sad state, they are more likely to remember unhappy memories from their childhood. Moreover, after reading about people who die from cancer, people estimate the probability that they will die from cancer higher than those who do not read about such cases. More interesting, however, is that they believe it more likely that they will die in an automobile accident, by gun shot, and other things. It seems that simply learning about one negative event increases the negative judgment about a wide range of such events.

FAMILIARITY AND HARMONY

Finally, there is an interesting result concerning the role of familiarity. In general, the more familiar something is, say, music, the more one likes it—up to a point. (c.f. Zajonc[5]). Eventually, a song, for example, can become boring. Our model provides an explanation for this phenomena. For example, suppose that one experiences a particular stimulus (say, a song) in a reasonably constant environment. Every time the song is heard, it is the occasion for a learning experience and the harmony will greater.

$$\mathcal{H} = g \sum_{ij} \left(\sum_k w_{ijk}\alpha_k \right) a_i a_j + \sum_i \beta i \qquad (4)$$

This is true because the weights will continue to grow on every learning trial. However, after long enough another factor comes in. The song will become increasingly predictable and therefore less surprising. As the song becomes less surprising, the arousal level, represented in the model by the "gain" term g, will become smaller and the harmony will be reduced. This suggests that the highest harmony will occur when the song, let's say, is familiar, but has a small variation which will lead to some unpredictability and therefore some arousal.

REFERENCES

1. Bower, G. H. "Mood and Memory." *Am. Psychol.* **36(2)** (1981).
2. Bower, G. H. "Affect and Cognition." *Phil. Trans. R. Soc. Lond. B* **302** (1983): 387–402.
3. Cohen, J., and D. Servin-Schreiber. "Context, Cortex and Dopamine: A Connectionist Approach to Behavior and Biology in Schizophrenia." *Psych. Rev.* (1991).
4. Zajonc, R. B. "Attitudinal Effects of Mere Exposure." *J. Personal. & Social Psych.* (Monograph Suppl.) **9** (1968): 1–27.

John Antrobus
City College of New York, Department of Psychology, New York, NY 10031

Thinking Away and Ahead

The ability to anticipate and plan, to imagine and think, even to dream, presumes the ability to construct, off line as it were—internal representations of events that are not present. Off-line processes may represent events that are removed from the present by large intervals of time or space, or they may consist of a succession of small steps away from the present.

Indeed, the process of perception—the process of perceiving the apparent "present"—is, paradoxically, a process of estimating a representation of the future from the evidence of the past. It takes at least 7 ms for an external stimulus to reach the brain, and another 300–700 ms for different portions of the brain to recognize it. In a dynamic world, therefore—imagine batting for the Yankees—the brain must learn to perceive and to respond to what will be, not what was when the stimulus enters the sensory pathway.

I suggest that as we move through space and time we continuously construct schemata of our future world—of what is next. In the large scale of events, the value to the individual of constructing such schemata is to minimize threat and maximize gain for the individual. Within the confines of the laboratory, this value may be realized by minimizing the error or the latency of perceptual and motor responses to stimuli. By computing the anticipated microfeatures, features, names,

and meanings of forthcoming stimuli the brain minimizes, the final computations required to recognize and respond to those stimuli when they do arrive.

In familiar environments, the process may be so highly learned, so automatized, that it produces no conscious evidence—no explicit hypotheses. Indeed, we may sometimes ignore critical sensory input and construct our schemata entirely from the prior context. Remember summer camp, when someone got out of bed at night, after switching from the lower to upper bunk, and made a crash landing.

Both the peripheral and central nervous system have a variety of devices to alert us when the environment becomes unpredictable, that is, when something does not fit this moving schema. For example,[4,34] changing the value of any high probability feature, such as the pitch or loudness of a repeating tone, will elicit an automatic cortical mismatch negativity response at about 100 ms. And changing any feature in a feature set, where the probability of the feature is highly dependent on the feature context, even where the context is cross-modal, will also elicit a mismatch negativity response. For example, if the bilabial burst that produces the sound /b/ or /p/ is replaced by /sh/, the cortex will produce a mismatch response.[42] The response occurs without our directed attention to the stimulus channel.

Once we do attend to the source, however, subsequent feature changes will elicit a second, slower cortical P3 response which has a latency of 200–600 ms. At a higher level of cortical processing, a word whose meaning or syntax is unexpected, for example, "He took a sip from the transmitter," elicits a cortical N400 wave.[11]

That the brain makes a unique response to any change in its input is not sufficient evidence that it uses its knowledge of the stochastic relations in the environment to anticipate future events. Evidence for such an anticipatory process is provided by parietal neurons that construct the expected target image of a saccadic eye movement. Lateral intraparietal neurons anticipate by about 150 ms, the retinal consequences of an intended eye movement. They remap the coordinates of the initial fixation onto those of the intended fixation—which then matches the reafferent visual input.[13]

Behavioral evidence of a more generalized kind is provided by verbal and visual priming experiments where the response latency for recognizing a word or visual object is shortened when it can be predicted from its preceding context.[50]

I have drawn attention to the forward-looking orientation of on-line, perceptual processes so that as I go on to describe off-line cognitive processes, I can point out some of the neural structures, and computational and adaptive functions shared by the two processes. Furthermore, we know far more about the first few milliseconds of perception than about all other cognitive processes. Our most precise models and empirical research describe how external stimuli initiate internal processes that produce behavioral responses in fractions of a second. It is only reasonable that we exploit these models in the attempt to understand off-line processes.

I would like now to show how (1) off-line and perceptual processes share certain cognitive and neural structures, (2) to describe the conditions under which they compete for those resources, and (3) to make some suggestions about the functional input for these "off-line" processes. I will then describe two neural network models

of some characteristics of off-line processes, and suggest how they may be used to guide future research.

ON- AND OFF-LINE IMAGERY AND THOUGHT: SHARED BRAIN PROCESSES

I will begin with the assumption that some of the neurocognitive processes of perception, particularly of novel events, are also utilized in the production of off-line cognitive processes. The assumption is supported by (1) neurophysiological evidence that similar brain regions are activated by on- and off-line processing of similar sensory features, (2) by behavioral evidence that when on- and off-line processes occur simultaneously, they interfere with one another, and (3) by behavioral evidence that the two processes normally alternate with one another. It is this alternation that justifies the dichotomous classification on- and off-line.

With respect to the participation of similar brain regions in on- and off-line activity, Roland and Friberg[38] showed that metabolism in the temporal and occipital cortices was elevated during auditory and visual imagery tasks, respectively. Evidence for similarity in single unit activity in the brain of the monkey in on- and off-line tasks was provided by Georgopoulos et al.,[18] when they demonstrated the successive activation of directionally tuned motor neurons in a mental rotation task. Similarly, magnetoencephalograms have been recorded in the human visual cortex during visual imagery.[28]

Behavioral evidence that visual imagery interferes with visual perception was first demonstrated in 1910 by Perky who showed that when subjects were instructed to form a visual image of an object, the threshold for recognizing an external perceptual stimulus increased. If a visual image increases the threshold for visual perception, then the two processes must share some common structures. This work was continued by Segal[44] at City College and recently confirmed in an excellent review by Craver-Lemley and Reeves.[9] Most of these studies examined images that were elicited by verbal instructions and they showed that imagery interfered with perception. Antrobus, Singer, Goldstein, and Fortgang[6] examined interference in the reverse direction. Studying spontaneously generated images, we found that the interference with mental imagery by a perceptual task was greatest where the sensory modality of a concurrent task and the imagery were the same.

Inasmuch as perception and mental imagery utilize some similar brain processes, mutual interference can be avoided only if the two processes alternate or switch back and forth. If the two processes are in competition for the same neural resources, one may assume that any switching process is controlled by a cost-benefit calculus that takes account of such factors as the adaptive significance of new sensory data, planning for anticipated gains, avoiding anticipated loss, and so forth. One limiting condition in this equation occurs when the sensory stimuli are repetitive and therefore highly predictable. This is the on-line condition described earlier where the environment is so redundant that the process of projecting the schema through space and time requires little or no attentional. At this point the risk, or expected cost, of switching off-line is at a minimum.

TUIT RESEARCH

In the early 1960s, Jerry Singer and I decided to study spontaneous imagery and thought in the laboratory. We assumed that private subjective experience was part of an informal cognitive process by which memories of personal concerns were activated in order to better enable an individual to make sense of the past, and to more effectively anticipate the future. We needed a rate measure that would be sensitive to the magnitude or strength of the neurocognitive processes that generate off-line imagery. In order to make it minimally obtrusive, we required that subjects make a simple judgement about the occurrence of a class of off-line cognitive events that we called Task Unrelated Imagery and Thought (TUIT). A TUIT is a conscious cognitive event that is unrelated to any concurrent external stimulus that has occurred within the trial interval. That is, the imaged features of the TUIT must lie outside of the experimental cubicle in space, and outside of the trial interval in time. The subject flipped a toggle switch at the end of each trial to indicate whether at least one TUIT, occurred within the trial, normally 15 seconds in duration.

In a sound-attenuated black cubicle, which is a maximally redundant environment, with no external information to process, subjects produce TUIT continuously. By adding a difficult sensory-motor task, however, we were able to push p(TUIT) down into a range where it could be used as a dependent variable to study the effect of the presentation rate of task information. As expected, p(TUIT) dropped as a linear function of task information rate—but only to a point. As subjects reached the upper limit of their ability to encode, p(TUIT) reached its lower limit of 0.35. Whether or not the brain produced off-line imagery was a function, therefore, of its rate of encoding the task information, and only indirectly a function of the information presentation rate. In subsequent studies we found that switching attention from the task to TUIT did not require long intervals of repetitive stimulation. In one experiment, increments in the interstimulus interval as small as 125 ms were sufficient to observe an increase in p(TUIT). The general picture we formed was that the brain typically runs off-line, except during those momentary intervals when its attentive processes are require to respond to novel or personally relevant stimuli.

If stimulus redundancy increases the likelihood of switching away from an external stimulus source, it must also determine the rate at which the brain switches from one off-line subject to another. This hypothesis was supported by an experiment in which we asked subjects, at unpredictable times within each one hour session over 11 days, to describe the content of their TUIT. Content was initially concerned with the most novel events in their environment, namely, the purpose of the experiment, the personal attitudes of the experimenter, and how the equipment worked, and then it gradually shifted away to matters of personal significance in their home, personal, and academic life.

In summary, the TUIT research supported the assumption that switching from on- to off-line processing, and within the latter, switching from one concern to another may be controlled by some cost-benefit process. Since the benefit of responding to a redundant source is close to zero, and the cost of switching away is close to zero, there is every reason for the focus of imagery and thought to go off-line in repetitive environments and, when off-line, to keep on changing.

BRAIN STEM AND THALAMIC CONTROL OF OFF-LINE PROCESSING

Without diffuse activation of the cortex, no mental experience can occur. This condition is most closely approximated in nonrapid eye movement sleep (NREM). By contrast, REM sleep, where most dreaming occurs, is characterized by an activated cortex. Identifying the subcortical neurophysiological processes that support the perception-like, but sensory-independent output of this brain state may help us understand how off-line mentation is produced in the waking state.

The pattern and magnitude of activation in the midbrain reticular formation, thalamus, and cerebral cortex is broadly similar to that of waking.[22,33,47,48] The brain looks awake, but the individual is behaviorally asleep because the sensory input and skeletal motor system are powerfully inhibited, or disfacilitated, at the level of the brain stem.[23] If the motor inhibition is surgically removed, the animal runs about chasing the hallucinated objects of its imagination.[29]

Despite the absence of retinal input, however, the brain continuously produces images that are sometimes as bright and clear as waking percepts. Using Steven's magnitude estimation procedure to scale the brightness and clarity of a set of color photographs, and asking subjects awakened from REM sleep to pick the photo most like different images they have just experienced, we found that the mean brightness and clarity of REM images are close to that of waking percepts, approximately 0.8.

Elsewhere[4] I have argued that image brightness and clarity and the estimated information in the mentation narrative in sleep are solely a function of the generalized or diffuse cortical activation that originates in the pons. Unfortunately it is almost impossible to obtain a satisfactory estimate of cortical activation during sleep. A useful indirect estimate, however, is provided by the fact that cortical activation varies spontaneously as a function of the phase of the 24-hour diurnal core body temperature rhythm and, in sleep, with the 90-minute REM-NREM rhythm.

In order to test this diffuse cortical activation hypothesis, we[5] recently delayed the sleep interval by three hours so that subjects would sleep through a point where both rhythms were in an activated phase. As expected, this sleep interval, i.e., very late morning, yielded unusually bright and clear visual images, especially so in late REM sleep. The sum of the two effects was linear. There were no differences in the pattern of imaginal attributes associated with the activated phases of the two rhythms. This means that the rising phase of the diurnal rhythm supports the same pattern of cortical activation as does REM sleep. What they share in common is the diffuse activation of the cortex, particularly the visual and spatial cortex.

In addition to the widely distributed diffuse cortical activation that supports off-line imagery and thought, both in REM sleep and the waking state, activation moves from one local cortical region to another in response to the changing characteristics of an individual's off-line thought.[38] Indeed, cortical energy consumption is greater during thinking than during active perception or physical exercise. The high ratio of cortico-thalamic fibers, in some areas 10:1, suggests that the cortex exercises precise local control over thalamo-cortical activation.

LaBerge[30] shows that when an individual expects to process a particular class of visual features, the appropriate cortical region displays increased metabolism, the index of increased regional activation. This expectation process is generated by what he calls "a thalamic enhancement circuit, which in turn is driven by a cognitive procedure,..." and the "selection of a particular cognitive procedure is assumed to depend on the momentary relative motivational value associated with it by means of connections to deeper limbic structures" (p. 358). LaBerge shows how the ability of cells in one cortical region to initiate regional activation in another provides a neural basis for how "considering something" might "start with a glimmer and then increase rather quickly to a full-blown entertainment of that 'something' that endures for a time. A 'clear'...consideration of an item would seem also to require the diminution of neural activities corresponding to similar (perhaps synaptically neighboring) items" (p. 366).

IMAGERY AND THOUGHT AS OFF-LINE INTERPRETATION

How does the brain that has learned perceptual sequences in response to external stimuli now produce the perceptlike sequences we call imagery and thought without the support of this sensory input? Because of its high sensory thresholds, some aspects of off-line imagery production can be examined most clearly in REM sleep. Imagined people and objects are "seen" just as though they are perceptions constructed from external stimuli. But what determines the construction of a particular image?

People who attempt to interpret the meaning of dreams implicitly assume that the images are created under the direction of some higher cognitive structure that somehow instructs the sensory projection areas to create visual images appropriate to its meanings and intentions. And so, interpreters of dreams use the image sequences to infer back to the intentions and meanings that supposedly initiated the dream.

But a dreamer often reports persons, objects, or features that s/he simply cannot identify, or says, after a pause, "....it was something like a..." If the features of these unrecognized images were created under constraints imposed by a high-level cognitive process, they surely would be, in effect, identified before they were produced.

Since this is often not the case, I have proposed elsewhere[4] that REM sleep images are constructed by the same processes as are waking percepts. Since the optic nerve is inhibited in REM sleep,[37] visual microfeatures from the lateral geniculate nucleus or primary visual cortex must pass through to higher cortical centers in a process of successive interpretation, where the imaginal output computed at each stage is indistinguishable from the real thing, i.e., a retinal pattern. According to Joseph Gleick,[19] Richard Feynman proposed a similar idea in an undergraduate philosophy paper at MIT about 1937. He suggested that the brain has an "interpretation department" that creates dream images from splotches of color that enter the eye. He had the makings of a great neuropsychologist.

Support for this interpretation notion comes from experimental studies of the "incorporation" of features of an external stimulus into an ongoing dream. They suggest that the context of the dream exerts powerful constraints on the interpretation of sensory features. Two examples from Bill Dement's Ph.D. dissertation[12] in physiology illustrate the point. Thirty seconds after the subject in REM sleep was sprayed with a fine water mist, he reported, "I was walking behind the leading lady, when she suddenly collapsed and water was dripping on her. I ran over to her and felt water dripping on my back and head. The roof was leaking. I was very puzzled why she fell down and decided some plaster must have fallen on her. I looked up and there was a hole in the roof. I dragged her over to the side of the stage and began pulling the curtains." The next subject was sprayed on his feet and legs while lying on his back during a very hot night in the lab. "Two children came into the room and came over to me asking for water. I had a glass of ice water and I tipped the glass to give it to them. I was sitting and I spilled the water on myself. The children wanted the ice and started grabbing it, but it slipped away. I got mad because they were so greedy and tried to shoved them away from the chair. Then I got out of the chair and was going to change my pants. As I left the room I seemed to be in a school and I saw lots of children in the hall and I seemed to be late for class. . ." (pp. 21–22).

Note the remarkable facility with which the brain interprets a stimulus that is completely out of context with the prior imagined thought. Notice also that the second subject's report of a sudden change of scene, "I seemed to be in a school. . .," implies the generation and interpretation of a new, internally generated visual image, but the phrase "I seemed to be late for a class. . ." appears to represent an intentional state that is constrained schematically by the visual image.

Further evidence for the interpretation hypothesis is suggested by anecdotes where the dreamer comments in the dream on the unexpected visual form. One dreamer remarked that his daughter's fingernail was flat, so he wondered whether she had banged it. Another dreamer described a table as having an odd shape, so she concluded that they must be at a circus.

As you can see, Feynman's interpretation department is a highly skilled one. Like the left hemisphere's interpretation of right-hemisphere-controlled behavior in the waking state, which are, in fact, fabricated explanations,[17] and the vast research literature on visual illusions such as Ames' window, the left hemisphere in REM sleep can produce a plausible account of just about anything.[2]

I say "just about" because subjects often throw in qualifiers such as "seemed like a ..." that suggest that the visual form is not perfect, or transitions such as, "and all of a sudden ... " that imply that the new person or object was out of temporal context.

Note that despite claims by Hobson and McCarley[23] and Crick and Mitchison (1986), there is no evidence that these improbable forms and sequences are associated with PGO waves. In fact, sudden shifts in topic are even more common among waking subjects when they are situated in the same quiet dark environment for a similar interval of time.[51]

In summary, I have suggested that the sequence of imagery and thought produced in one cortical region during REM sleep is a function, first, of diffuse cortical activation and high sensory thresholds[23]; second, of the memorial context of its own prior activity; and, third, perhaps intermittently, of input from cortical regions that are dedicated to a particular sensory modality, particularly the visual modality, and whose output is independent of the processes of the cortical region where the final interpretation takes place.

How to describe the memory process that enables similar but different patterns of sensation to be transformed into different perceptions, is exactly what Sir Henry Head[7] struggled with in 1920 as he studied the inability of patients with parietal lesions, with their eyes closed, using only tactile information, to perceive limb position. What is the form of the tactile-visual memory that makes this spatial discrimination possible for intact individuals? Head rejected the notion that memory consists of a list of unique traces of specific events, as well as the alternative concept, that memory is some representation of central tendency—because neither alternative was sufficiently flexible to account for the unique perceptual responses that subjects make to unique tactile patterns. He proposed instead the term schema to imply a memory that has some intermediate generalized form.

CONSTRAINT SATISFACTION IN IMAGERY AND THOUGHT

With the publication of *Parallel Distributed Processes* in 1986, Rumelhart and McClelland and their colleagues proposed, as you know, that memories may be represented as patterns of connections among distributed neuron-like or microfeature units. Rumelhart, Smolensky, McClelland, and Hinton[41] proposed a model of perception in which the activation of a few of these units by an external source, initiates the propagation of activation throughout the network that continues until it settles onto a pattern of activation that best satisfies the joint constraints imposed by the input and the previously learned connections among the microfeatural units in the network. Satisfying the joint constraints of input and memory is a process by which the microfeatural units self-organize to create a coherent representation of the original stimulus.

One attractive feature of this schema model is its ability to simultaneously represent different levels of abstraction. For example, in the process of constraint satisfaction, units representing small sets of closely related microfeatures will cluster together to form a subschema, which, in turn, will then behave as though it were a single unit in the larger network. Several such subschemata will then behave like macrofeatures and cluster together with other closely related subschemata to form a single schema.

The ability of the model to simultaneously satisfy both external and memorial constraints makes it a powerful mathematical tool for representing as an emergent characteristic, the subjective sense of the unity of conscious experience. As you can see, this constraint satisfaction process is exactly what, in the context of dreaming, we have called interpretation.

But once the network satisfies its constraints, it stops until it receives new input. If it is to represent off-line processing, it must therefore get input from some other location in the brain.

Rumelhart and his colleagues proposed an extension of the schema model in which a schema network that can ask questions about the world is linked to an independent network representing world knowledge. In response to each question the world network can compute an answer in the form of a schema, and each answer can initiate a constraint satisfaction process that computes a new question, in the form of a new schema. In this manner the model can propagate through time. As far as I know, no one has yet developed a working simulation of this model. In its broad outline, however, it suggest ways in which models of a schema, or a pair of schemata, might represent an interior monologue or a visuo-motor sequence that moves through imagined space and time.

DUAL AND UNIMODALITY IMAGINAL SEQUENCES

In order to identify the processes by which temporal sequences of imagery and thought can be produced, it is useful to distinguish processes that are executed as interactions between two modalities, such as imagined speaking and imagined hearing, and sequences propagated within a single modality. Many of our waking perceptions are the result of motor processes that we ourselves initiate—what Skinner called response-produced stimuli. Responding to the first stimulus, the door, you open it, then you see the second stimulus—who's inside. Skinner used these chains to account for well-learned sequences such as tying one's shoes or starting one's car. The succession of visual stimuli is determined not by sequences in the external environment but by the behavior of the individual.

Extended sequences of imagined responses followed by imagined response-produced stimuli are rather typical of the interior monologues described by the "double agent" model[31,46] recently reviewed critically by MacKay.[32]

REM sleep mentation seems to include both dual and unimodality imagery, but very little of it includes speech. Typical of the dual modalities in REM sleep is a visual image followed by an imagined head and eye movement, followed in turn, by a new image. Although no one has studied these sequences systematically, I suspect the new image is more often than not unexpected because it is jointly constrained, not only by the contextual expectation, but by and whatever pattern is independently provided by the primary visual cortex. This unexpected quality gives the dream a real-life quality that it could not have if it were driven by a single top-down source.

From this viewpoint, the question-and-answer schema model proposed by Rumelhart et al. is an example of a dual modality model. It propagates through time by passing information back and forth between two networks, each representing different cortical regions.

There is evidence that sequences of imagery can also propagate forward within a single cognitive modality. For example, subjects awakened from REM sleep often liken their imagery sequences to watching TV or a movie. The dreamer does not participate. Howard, Mutter, and Howard[27] have shown recently that serial visual pattern learning can be achieved without any intervening motor response. For example, if you are sitting in a room, and the door opens, you expect to see someone walk in, and when they do you expect them to acknowledge you. After that point, the predictability of what they do next may be so low, that some nonsequential associative process will determine the next image, at which point a new sequence may begin.[3]

SIMULATION OF DREAMING WITH THE SCHEMATA MODEL

In order to simulate an off-line unimodal process as in dreaming, a network must be able to run indefinitely without concurrent external input. If there is input, such as visual microfeatures from some other region of the brain, it is unlikely to be sustained throughout the constraint satisfaction process as it is in the Rumelhart et al. model. Even in waking perception, the cortex does not tend to sustain a representation of sensory input.[36] But if the input to the schema network is not clamped, it is lost and the network computes a schema that is simply a characteristic of its overall dominant weight pattern.

O UNIT

o WEIGHT

<-- ACTIVATION   OR   INHIBITION

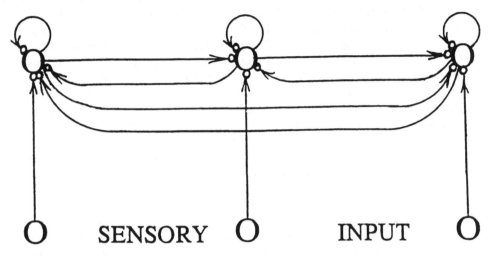

SENSORY   INPUT

FIGURE 1   Single-layered schema network.

I found that by adding a slow decay loop to each unit of the network, I could replace the clamping operation with only a single activation of one or two units— and the network would compute quite satisfactory schemata (see Figure 1). But unlike real off-line thought and imagery, of course, this slightly modified model still computed to a solution and then stopped. Furthermore, once the hill-climbing process was well underway, if the network was very large the schema model was unresponsive to new input. In this sense it could not simulate the incorporation of new external stimulus features into an ongoing image sequence.

SIMULATING TEMPORAL SEQUENCES IN A SINGLE MODALITY WITH A SINGLE-LAYERED NETWORK

How can a single-layered network compute from one state to another, where each state represents an ordered sequence in time? Based on Hopfield's[25] simulated annealing model, Rumelhart's schema model assumes that the units of the network are connected by symmetrical weights ($w_{ji} = w_{ij}$). But in order for a network to compute sequentially from one schema to another, its weights must be asymmetric, and such networks are known to exhibit oscillation and chaos.[26] Human off-line imagery and thought may well be chaotic, of course, but until we find a way of measuring off-line mentation without totally interrupting the process we may never know how chaotic.

Despite its vulnerability to the disorganizing consequences of chaos, a network with asymmetric weights seemed like a plausible way to represent the production of the off-line imagery and thought of REM sleep. So I built a simple network that could move from one schemalike cluster of activated units to another. All units were assigned a moderate time constant. Units within each cluster were linked with high positive weights. They were linked to units in the next cluster with smaller positive weights, and to a prior cluster of units with inhibitory weights. When any unit, in any cluster, was activated, activation spread to the entire cluster of units forming a schemalike organization as in the Rumelhart model. But instead of stopping, the cluster of active units passed activation to the units in the next cluster, which when active, inhibited the units in the first schema.

With three or more such clusters of units, arranged so that the last activates the first, the network can propagate forever without becoming disordered—though I did not measure the chaos in these simulations. If, however, any unit in the network participates in two or more clusters, the network quickly becomes disordered. How the biological brain solves this problem I do not know. But it has at least two procedures for reducing the associative spread of activation throughout an asymmetrically weighted network. One is based on the response characteristic of individual neurons. For example, information in the sensory pathways is controlled by different neurotransmitter, and their neurons have shorter time constants than do associative neurons.[49] In my simulations of sequences in the life of a college freshman, I assigned the longest time constants to typical roles such as eating, playing ball, and studying, and short time constants to visual features. In this way the role units, which, of course, were mutually inhibitory, tended to dampen irrelevant activity in the units with shorter time constants.

Chaos may also be minimized by limiting the proportion of neurons that can become active at any one time. The limiting process may be implemented by cortico-thalamo-cortical circuits which can moderate the size of the cortical region that receives diffuse activation. This process may be simulated with Sigma-Pi units, so that the activation of each cortical unit is dependent on the product of regional and local activation.

The use of Sigma-Pi units to simulate these cortico-thalamo-cortical circuits also provides a basis for simulating some of the differences between waking and REM sleep off-line mentation. The inhibition, or disfacilitation,[24] of connections between the commands of the motor cortex and the skeletal muscles, between proprioceptive sensors and the sensory cortex, is unique to REM sleep. The motor cortex continues to give commands to the motor system, but the expected feedback never arrives. The dreamer's response is "I'm paralyzed." The simulation of this sequence is accomplished rather easily if the distribution of diffuse activation to different regions of the schema network—(1) thought and visual imagery, (2) motor intention and commands, and (3) sensation including proprioception—is controlled by a unit that represents the diurnal wake-sleep, and the REM-NREM sleep cycles.

Although some features of the "I'm paralyzed" sequence would be much easier to simulate with a script model,[45] this asymmetric schema model still enjoys all the advantages of a parallel distributed network. For example, priming the network with a single feature, such as ball, is sufficient to generate a sequence of schemata that represent an entire game of catch.

But whatever the merits of this model, it cannot represent nonlinear functions. For example, a spray of water cannot represent a leaky roof in the context of standing backstage in a theater, as well as a spilled glass of water in one's living room. In a linear model, the joint effect of any set of units is always the sum of their separate effects.

MODIFIED BACK PROPAGATION: NONLINEAR SEQUENCES WITH LEARNED WEIGHTS

With its hidden layer of units, the output layer of the back-propagation model[40] can express output patterns that are nonlinear functions of input patterns, in particular, contextual relationships represented by input units. Moreover, the architecture of the back-propagation model may be modified so as to represent sequential relationships that are characteristic of off-line sequences. In so doing, however, the back-propagation model preserves only some of the spirit, and none of the formal characteristics of the single-layered schema model. Unlike the schema model, where each unit has reciprocal connections to each other, the standard back-propagation model passes a pattern of activation from an input layer through an intermediate or hidden layer and terminates at the output layer. The pattern produced at the output layer is the "best" response to the input pattern in the sense that it is determined by sets of weights that have been learned from past experience with a large set of input-output patterns. Unlike the schema model where the weights are based on the probability of the co-occurrence of individual item pairs, the weight structure in back propagation is based on the probabilities of the co-occurrence of pairs of entire patterns of units. In this manner, back propagation can represent nonlinear contextual relations between the units in the input and output patterns.

Back-propagation models demonstrate stimulus generalization. Presented with a stimulus that is similar to one for which it has learned to produce a particular response, the network will produce that response. If the stimulus is quite novel, however, the response may be ambiguous or degraded. Unlike the continuous hill-climbing process that the schema model uses to compute a unique solution to a unique input, back prop tends to produce the best response in its learned repertoire.

Now let us consider whether back propagation can simulate the production of off-line event sequences within a single sensory modality, and whether it can simulate the incorporation of novel, perhaps degraded input such as I have suggested might be provided by the visual pathways during REM sleep.

In order to simulate this process, I described a sequence of images, where each image consisted of five binary digits. The sequence was learned by giving the network the first image as input, and training it to produce the second as output; giving it the second as input, and training it to produce the third; and so forth. I was particularly interested in whether, in its off-line, or imaginal, mode, which I will describe shortly, the network would become progressively disordered. So I trained the 28th image to produce the first, so that in the off-line image mode, the sequence could loop indefinitely. My concern about progressive disorder had been raised when I found that every large, single-layered, sparsely connected network with asymmetric weights appeared to become disordered after about ten iterations of the network. In nets of more that 150 units, with sparse connections, individual clusters of units tended to develop a life of their own.

In order to simulate a context effect, the 28 sequences were divided into 4 subsequences, each with 7 sequences. The contexts were expressed by the first two units in each image; the remaining three units, nested within each context described a sequence across the seven images. The sequences differed within each context. The network was then trained to a least-squares error criterion of 0.01 for each unit, in each image.

Simulation of the off-line or imagery mode consisted of returning the output of the network back to the input layer (see Figure 2). This return to the input of its own output is based on the simple model that in the waking perception mode, thalamic activation of the input pathways inhibits the return of the output of the perceptual areas of the cortex to their input. In REM sleep, the noradrenergic brain stem—thalamic facilitation of perceptual input is removed,[24] thereby allowing the output of this portion of the brain to pass back to the input.

Like the schema model where all units are directly and symmetrically connected, in this recurrent back prop model, all units are reciprocally connected to each other, but the connections between units are indirect, mediated by differently weighted connections to each unit in the hidden layer. And, of course, each hidden unit also receives activation from all other units in the input layer, so that each of the connections between any pair of units in the input layer can be modified by the contextual activation of the other input units. As in the single-layered net, the connections between input units are reciprocated but they are almost never symmetrical. It is these multiple representations of the input units by the hidden units that permit the model to discriminate nonlinear patterns among the input units such as contextual or conditional relations.

When, in this off-line, imaginal, or recurrent mode, the network is primed with any of the 28 images in the training set, it propagates forward as expected and produces the next image from the original training set, and then the next and

O UNIT
o WEIGHT
←─ACTIVATION
⊢ INHIBITION

HIDDEN

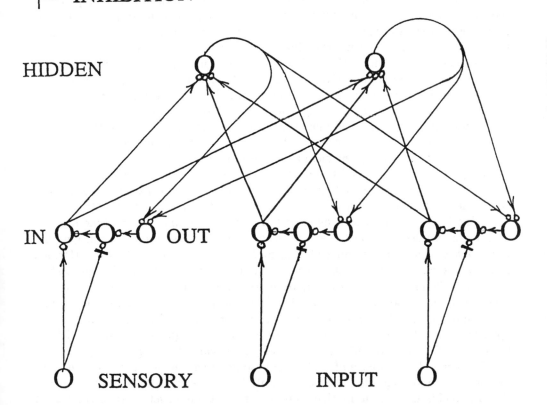

IN   OUT

SENSORY   INPUT

FIGURE 2 Back-propagation network with the output layer returned to the input layer, except when inhibited by sensory input.

so forth. Note that while the initial input is expressed in binary form, the recurrent input from the output layer is expressed in continuous values. Unlike the schema model, this model shows no tendency to become progressively disordered.

As you recall from the so-called stimulus incorporation dream studies, a spray of water, or other external stimulus that is out of context with respect to an ongoing dream will be interpreted in a manner that is appropriate to the context of the dream. In its off-line mode, back prop also produces context appropriate responses. When primed with a novel image—where one or more of the last 3 units is set to

0.5, rather than 0 or 1—or primed with an image that is a combination of units that did not occur in the training set, the network produces, as its next image, one of the images that it had previously learned, and that image is always contextually appropriate with respect to the context units. Although the first response to the novel image tends to be somewhat degraded, it is typically sufficiently good for the net to continue forward with the learned sequence.

In simulations where, in the training mode, the learning of one step in the sequence barely met the minimum learning criterion, the same step tended to be even worse in the off-line simulation. The poorer performance was due, of course, to the error in approximating the binary coded input of the training set with the continuous-valued input of the off-line mode. In a few simulations where, in the training mode, the learning of one sequence fell below the minimum criterion, the added error in the input image in the off-line mode led to near random values for one or two units in the output image. But, to my surprise, as the network continued to compute its sequence of off-line output, within two to four cycles, the network was again producing good output. Although the first good item tended to be out of order, all of the subsequent members were in sequence.[4]

Degraded images tended to be degraded in the last three units of the image rather than the first two context units. Recovery from a degraded image step in the sequence, therefore, invariably occurred within the same context as that of the degraded image. That was to be expected because the first two context units in each image tended to repeat themselves, whereas the three sequence units within each context changed from image to image. In this respect, the network performed rather like someone who has learned a tune imperfectly. S/he may get it right up to a point, and then fail to get the next note. At that point, after some rather unmusical sound, s/he may loop back to some better learned, but contextually appropriate point, and continue. I suggest, however, that not only the well-learned, but even the apparently novel sequences of off-line imagery and thought are constructed from fragments—features and microfeatures—of highly learned perceptual sequences.

The generalizing ability of back prop is so good that it produced good-quality responses to both degraded and novel input. But it never produced novel images, that is, combinations of units that were not in the training set. Part of the reason for the lack of novel output was that of the 32 (2^5) possible output images in the training set, 28 were employed in the output set, so the probability of a novel image was only 0.125. It was as though almost everything one sees belongs to some previously learned category.

I tested this hunch by increasing the number of units per item so that the expected probability of a novel image was 0.75. Now the probability of a novel image increased to about 0.04. As before, the image was always appropriate to the context of the input image. And as with the moderately degraded images in the previous simulations, the novel image always propagated forward to an image from the original training set.

The remarkable generalizing ability of back prop to render as output an image that is similar to one that it has previously learned, nicely represents the interpretive characteristic of off-line mental processes. It demonstrates the ability of the net to impose order on the output, even where little was present in the input. Further, it demonstrates that the learned nonlinear contextual relations between input and output patterns can determine this order.

As one may see, the sequential constraints in this model are confined to those learned in the simulated waking perception mode. Other than the loop back from the output to the input layer, the model does not have an independent memorial record of its off-line or imaginal history. Jordan,[21] Cleermans, Sevan-Schreiber and McClelland,[8] and Elman[14] have studied simple recurrent back prop models of perceptual processes, but how they behave when their output is looped back to their input has not yet been studied. Interaction between the two recurrent loops would undoubtedly produce hysteresis. But the models should be studied.

In conclusion, I have drawn your attention to the substantial amount of time that the brain devotes to off-line processing. Although the quality of off-line imagery never quite matches that of waking perception, it is greatly underrated by current models of perception, many of which assume the brain is in a null state until provided with sensory input. And so I think that constructing better models of off-line processes will contribute to the understanding of perception.

The parallel distributed networks of Rumelhart and his colleagues, which I have modified here to represent some aspects of off-line processes, are powerful mathematical and conceptual tools that bring much needed clarity to our understanding of these processes. But as you can see, more work must be done to develop network architectures that change across time when the network is off-line—and in satisfying constraints within and between networks that represent activity in different brain regions. In this regard, I think there is merit in studying the collaboration of local constraint satisfaction processes carried out by cortical neurons, with constraints imposed by cholinergic, adrenergic, dopaminergic, and other neurotransmitter systems controlled by subcortical structures, and with constraints carried out by the cortico-thalamo-cortical enhancement circuits proposed by LaBerge.

REFERENCES

1. Antrobus, J. "Information Theory and Stimulus-Independent Thought." *Brit. J. Psych.* **59** (1968): 423–430.
2. Antrobus, J. "Cortical Hemisphere Asymmetry and Sleep Mentation." *Psych. Rev.* **94** (1987): 359–368.
3. Antrobus, J. "Dreaming: Cognitive Processes During Cortical Activation and High Afferent Thresholds." *Psych. Rev.* **98** (1991): 96–121.
4. Antrobus, J., S. Alankar, D. Deacon, and W. Ritter. "Auditory Orienting: Automatic Detection of Auditory Change Over Brief Intervals of Time: A

Neural Net Model of Evoked Brain Potentials." *Proceedings of the International Joint Conference on Neural Networks* **3** (1991): 775–779.

5. Antrobus, J., T. Kondo, R Reinsel, and G. Fein. "Dreaming in the Late Morning: Summation of REM and Diurnal Cortical Activation." *Consciousness & Cog.*: in press.

6. Antrobus, J. A., J. L. Singer, S. Goldstein, and M. Fortgang. "Mindwandering and Cognitive Structure." *Trans. NY Acad. Sci.* **32** (1970): 242–252.

7. Bartlett, F. C. *Remembering.* Cambridge: Cambridge University Press, 1938.

8. Cleermans, A., D. Servan-Schreiber, and J. L. McClelland. "Finite State Automata and Simple Recurrent Networks." *Neural Comp.* **1** (1989): 372–381.

9. Craver-Lemley, C., and A. Reeves. "How Visual Imagery Interferes with Vision." *Psych. Rev.* **99** (1992): 633–649.

10. Crick, F., and G. Mitchison. "REM Sleep and Neural Nets." *J. Mind & Behav.* **7** (1986): 229–250.

11. Deacon, D., and F. Breton. "The Relation Between N2 and N400: Scalp Distribution, Stimulus Probability, and Task Relevance." *Psychophysiology* **28** (1991): 185–200.

12. Dement, W. "The Physiology of Dreaming." Unpublished Ph.D. Thesis, Department of Physiology, University of Chicago, 1958.

13. Duhamel, J-R., C. L. Colby, and M. E. Goldberg. "The Updating of the Representation of Visual-Space in Parietal Cortex by Intended Eye Movements." *Science* **255** (1992): 90–92.

14. Elman, J. L. "Finding Structure in Time." *Cog. Sci.* **14** (1990): 179–211.

15. Farah, M. J.,and A. F. Smith. "Perceptual Interference and Facilitation with Auditory Images." *Percep. & Psychophys.* **33** (1983): 475–478.

16. Friedman, H. R., J. D. Janas, and P. S. Goldman-Rakic. "Enhancement of Metabolic Activity in the Diencephalon of Monkeys Performing Working Memory Tasks: A 2-Deoxyglucose Study in Behaving Rhesus Monkeys." *J. Cog. Neurosci.* **2** (1990): 18–31.

17. Gazzaniga, M. S., and J. E. LeDoux. *The Integrated Mind.* New York: Plenum, 1978.

18. Georgopoulos, A. P., J. T. Lurito, M. Petrides, A. B. Schwartz, and J. T. Massey. "Mental Rotation of the Neuronal Population Vector." *Science* **243** (1989): 234–236.

19. Gleick, J. *Genius: The Life and Science of Richard Feynman.* New York: Pantheon, 1992.

20. James, W. *The Principles of Psychology.* New York: Dover, 1890, 1950.

21. Jordan, M. I. "Attractor Dynamics and Parallelism in a Connectionist Sequential Machine." In *Proceedings of the Eighth Annual Conference of the Cognitive Science Society*, 531–546. Hillsdale, NJ: Erlbaum Associates, 1986.

22. Hobson, J. A. "The Effect of Chronic Brain Sem Lesions on Cortical and Muscular Activity During Sleep and Waking in the Cat." *Electroencep. & Clin. Neurophysiol.* **19** (1965): 41–62.

23. Hobson, J. A., and R. W. McCarley. "The Brain as a Dream State Generator: An Activation-Synthesis Hypothesis of the Dream Process." *Am. J. Psych.* **134** (1977): 1335–1348.

24. Hobson, J. A., and M. Steriade. " Neuronal Basis of Behavioral State Control." In *Handbook of Physiology*, edited by V. B. Mountcastle, F. E. Bloom, and F. E. Geiger, 701–823. Bethesda: American Physiological Society, 1986.

25. Hopfield, J. J. "Neural Networks and Physical Systems with Emergent Collective Computational Abilities." *Proc. Natl. Acad. Sci. USA* **79** (1982): 2554–2558.

26. Hopfield, J. J., and D. W. Tank. "Computing with Neural Circuits: A Model." *Science* **233** (1986): 625–633.

27. Howard, J. H. Jr., S. A. Mutter, and D. V. Howard. "Serial Pattern Learning by Event Observation." *J. Exper. Psy.: Learning, Memory and Cognition* **18** (1992): 1029–1039.

28. Kaufman, L., B. Schwartz, C. Salustri, and S. J. Williamson. "Modulation of Spontaneous Brain Activity During Mental Imagery." *J. Cog. Neurosci.* **2** (1990): 124–132.

29. Jouvet, M. "Neurophysiology of the Status of Sleep." *Phys. Rev.* **47** (1967): 117–177.

30. LaBerge, D. "Thalamic and Cortical Mechanisms of Attention Suggested by Recent Positron Emission Tomographic Experiments." *J. Cog. Neurosci.* **2** (1990): 358–372.

31. Levelt, W. J. M. *Speaking: From Intention to Articulation.* Cambridge, MA: MIT Press, 1989.

32. MacKay, D. G. "Constraints on Theories of Inner Speech." In *Auditory Imagery*, edited by D. Reisman, 121–149 . Hillsdale, NJ: Erlbaum Assoc, 1992.

33. Moruzzi, G., and H. W. Magoun. "Brain Stem Reticular Formation and Activation of the EEG." *Elec. & Clin. Neurophy.* **1** (1949): 455–473.

34. Naatanen, R. "The Role of Attention in Auditory Information Processing as Revealed by Event-Related Potentials and Other Brain Measures of Cognitive Function." *Behav. & Brain Sci.* **13** (1990): 210–288.

35. Perky, C. W. "An Experimental Study of Imagination." *Am. J. Psych.* **21** (1910): 422–452.

36. Phillips, D. P., R. A. Reale, and J. F. Brugge. In *The Neurobiology of Hearing*, edited by R. A. Altschuler, R. P. Bobbin, B. M. Clopton, and D. W. Hoffman, 335–365. New York: Raven, 1991.

37. Pompeiano. "Mechanisms of Sensorimotor Integration During Sleep." In *Progress in Physiological Psychology*, edited by E. Stellar and J. M. Spraque, Vol. 3. New York: Academic Press, 1970.

38. Roland, P. E., and L. Friberg. "Localization of Cortical Areas Activated by Thinking." *J. Neurophys.* **53** (1985): 1219–1243.

39. Rumelhart, D. E., G. E. Hinton, and J. L. McClelland. "A General Framework for Parallel Distributed Processing." In *Parallel Distributed Processing:*

Explorations in the Microstructure of Cognition, edited by D. E. Rumelhart and J. L. McClelland, Vol. 1, 45–109. Cambridge, MA: MIT Press, 1986

40. Rumelhart, D. E., G. E. Hinton, and R. J. Williams. "Learning Internal Representations by Error Propagation." In *Parallel Distributed Processing: Explorations in the Microstructure of Cognition*, edited by D. E. Rumelhart and J. L. McClelland, Vol. 1, 318–362. Cambridge, MA: MIT Press, 1986.

41. Rumelhart, D. E., P. Smolensky, J. L. McClelland, and G. E. Hinton. "Schemata and Sequential Thought Processes in PDP Models." In *Parallel Distributed Processing: Explorations in the Microstructure of Cognition*, edited by D. E. Rumelhart and J. L. McClelland, Vol. 2, 7–57. Cambridge, MA: MIT Press, 1986.

42. Sams, M., R. Aulanko, M. Hamalainen, R. Hari, O. Lounasmaa, S. Lu, and J. Simola. "Seeing Speech: Visual Information from Lip Movements Modifies Activity in the Human Auditory Cortex." *Neurosci. Lett.* **127** (1991): 141–145.

43. Schank, R. C., and R. P. Abelson. *Scripts, Plans, Goals, and Understanding* . Hillsdale, NJ: Lawrence Erlbaum, 1977.

44. Segal, S. *Imagery: Current Cognitive Approaches.* San Diego: Academic Press, 1971.

45. Shank and Abelson. *Scripts, Plans, Goals, and Understanding.* Hillsdale, NJ: Erlbaum Assoc., 1977.

46. Smith, J. D., D. Reisberg, and M. Wilson. "Subvocalization and Auditory Imagery: Interactions Between the Inner Ear and Inner Voice." In *Auditory Imagery*, edited by D. Reisman, 95–119. Hillsdale, NJ: Lawrence Erlbaum, 1992.

47. Steriade, M., N. Ropert, A. Kitsikis, and G. Oakson. "Ascending Activating Neuronal Networks in Midbrain Reticular Core and Related Rostral Systems." In *The Reticular Formation Revisited*, edited by J. A. Hobson and M. A. B. Brazier, 125–167. New York: Raven Press, 1980.

48. Steriade, M., K. Sakai, and M. Jouvet. "Bulbo-Thalamic Neurons Related to Thalamocortical Activation Processes During Paradoxical Sleep." *Exper. Brain Res.* **54** (1984): 463–475.

49. Tucker, D. M., and P. A. Williamson. "Asymmetric Neural Control Systems in Human Self-Regulation." *Psych. Rev.* **91** (1984): 185–215.

50. Tulving, E., and C. Gold. "Stimulus Information and Contextual Information as Determinants of Tachistoscopic Recognition of Words." *J. Exper. Psych.* **66** (1963): 319–327.

51. Wollman, M., and J. Antrobus. "Sleep and Waking Thought: Effects of External Stimulation." *Sleep* **9** (1986): 438–448.

Roger C. Schank and John B. Cleave
Institute for the Learning Sciences, Northwestern University, 1890 Maple Avenue, Suite 300, Evanston, IL 60201

Natural Learning, Natural Teaching: Changing Human Memory

The method people naturally employ to acquire knowledge is largely unsupported by traditional classroom practice. The human mind is better equipped to gather information about the world by operating within it than by reading about it, hearing lectures on it, or studying abstract models of it. To understand this, compare two alternative strategies someone might use for learning how to operate a VCR. In one, the person sets up the VCR, gets out a tape, and begins experimenting. He may reference the manual to find some specific piece of information, but he figures it out mainly by pressing buttons, watching for responses, and modifying his procedure accordingly. In the second strategy, the person pulls out the VCR's manual and begins studying it. He memorizes the functions of the various buttons, familiarizes himself with step-by-step instructions for loading/unloading the tape, recording, playing, etc., and learns general rules and cautions regarding the machine's use.

The first strategy is an instance of what we call "natural learning." We use the term "natural" because, given a choice between the two, most people would opt for the first, as they do for learning most of their skills—using computers, driving cars, dating, playing games, shopping, and so on. They choose to learn that way because it is most compatible with the way they maneuver through life.

On the other hand, much of today's teaching practice—focusing on vocabulary lists and taxonomies, teaching abstract rules, and presenting information apart from contexts in which it might be useful—conforms more to the approach embodied in the VCR manual. This is unfortunate, because teaching that goes against the grain of how people intuitively learn will not be successful. It only makes people come to dread school, for the same reason they come to dread computer manuals. Effective instruction augments natural learning, enabling students to employ strategies with which they have had much practice. The principles for such instruction rest on the process by which people naturally acquire knowledge about the world.

HUMAN MEMORY
THE REAL STORY

Human memory is story-based.[3] Stories are the way people package their experience into a coherent, functional framework, and everyone has a huge collection of them, concerning all aspects of their existence. People do not remember everything as a story, however. Stories usually involve exceptional situations they have been in: most people's memory of high school, for example, is captured in stories about salient events, such as their first date, acting in a play, wrecking the family car, or receiving an award. People remember these things long after they have forgotten the specific subject matter that they were formally taught at the same time.

The reason we remember stories so well is that they help us to make sense of our world and, thus, to get along better in the future. Remembering stories about previous trips to a country helps returning visitors avoid the same pitfalls, such as leaving exposed luggage in the airport or going to that restaurant with a particularly rude waiter. We comprehend the situation we are in by recalling ones we have experienced before, and using them to figure out what is going on. We can make our way through a bookstore or restaurant by recalling previous trips to bookstores or restaurants. In the course of searching our memory for relevant prior experiences, what we recall is often a story, which presents information in a form that is readily applicable, capturing details on negative consequences of actions, alternative plans, potential pitfalls, and other things to guard against.

We want others to profit from our experience as well, but the only way we can share it is through stories. When we want someone to understand some episode we were in, we usually must provide many details the person can use to make sense of it—the context, the causes and consequences of events, and the agents involved and their motives. We must present those details in such a way that the person can match them to their own encounters. The way we do this is by telling a story, a framework the listener recognizes and knows how to process. Every culture protects and preserves their customs in stories, and parents use them to advise their children. Conversation, in fact, consists largely of people trading stories. One person describes an event of which they were part, the other replies with a similar tale. We engage in such conversation every day, without much conscious effort. We do so not only for others' benefit, but our own as well; telling people about things that happen to us gells our own perspective on them. In order to be able to generalize from our own experiences, we must be able to communicate them, as stories, to others.

Stories are the basic unit of commerce in our collective mental life, the means by which we recognize that someone understands us. To see this, suppose John tells Mary about trouble he had getting his car repaired on vacation. To understand it, Mary must try to match it to her own experience. What she recalls, and then chooses to tell, indicates how well she understood John's story. If Mary replies with a story about her vacation otherwise unrelated to car repair, John might think she had missed his point. A story about being stranded with a flat on a cross-country trip perhaps shows a deeper understanding of John's predicament. Intelligence is all about knowing the right story to tell at the right time.

Since stories are the way we package and relate our experiences, in a form that makes it possible to apply them in new contexts and to relate them to each other, how we collect and recall them should be of central concern to educators. The process by which we do so can be crudely divided into three phases: the search, the expectation, and the failure.

THE SEARCH

How are people able to recall the right story to tell, at the time they need to tell it? To find a story, they must have indexed it beforehand, that is, attached some sort of labels to it so they can later remember it at an appropriate time. As people make sense of present experience by comparing it to past experiences, they index it in terms of that comparison. A man on the sidewalk dressed in a chicken outfit will seem odd unless someone can recall other incidents of people in costume soliciting people; when such a recognition is made, the person indexes the experience (i.e., remembers it later) as "another business promotion." Stories often have many indices, relating them to other stories and events in our memory. The more richly a memory is indexed, the more broadly it can be applied. Older people, especially, often have stories so richly indexed they seem applicable to almost anything. Since each of us has had a unique set of experiences, each has a unique set of indices as well, although we have many in common by virtue of the similarity of our daily life.

To construct an index for a story, we assess its content, selecting information useful in retrieving something similar in memory to shed light on it. In the case of the business promotion, above, a person might detect that it is a man wearing a large suit (and not a huge chicken), that the man stays in front of a restaurant (possibly ruling out a masquerade party), that the restaurant looks new, and that the man is handing out fliers (as opposed to, say, acting crazy). The information we gather about an experience we use to recall the incident later. The more information we collect, the more raw material for creating indices—the more "grist for the indexing mill," so to speak. When given a rule of thumb without a context, such as "honesty is the best policy," people have little with which to connect to their experience and, hence, little way of remembering it. Planted in a story, however (such as one where a person lies to his customers, gets caught at it, and so loses them), it is more memorable, because it has much more detail to use in indexing. When we become interested in something, we gather more information about it, reflect on it more, and, hence, connect it more richly to our other memories. Purchasing a home for the first time, to take one case, is often a salient memory for people, because of the amount of cognitive effort they expended. On the other hand, we let much of our experience go by without indexing it. It is difficult to distinguish one morning's breakfast from any other. It is hard to remember the shirt you wore two days ago.

The way that we match the situation we are in to those in our memory is by seeking content similarities. At one extreme, many situations are almost identical to ones previously encountered, in which case it takes little cognitive expenditure to find a match. For example, most people's drive to work on one morning is so similar to their drive to work on other mornings that they are able to execute it without thought: they use features of their situation (it is morning, they are in the kitchen, they have to get to work) to recall a very generalized plan (get in the car, drive north on Elm, etc.), and were able to get to work with little conscious effort. Similarly, people can easily negotiate a new restaurant, because restaurants tend to be very similar. At the other extreme are situations unlike any we have seen: these we must think about much more consciously and deliberately. Someone who encounters their boss at the mall for the first time, for example, can get flustered, because they do not know which of their previous experiences are applicable: should they treat him like they do at work, or like they would a familiar acquaintance? (Of course, the next time this happens, they have precedent to guide their actions.)

In short, the process we use to recall stories is part of the same process we use to store them. We make sense of present experience by comparing it to previous ones; once we have found a match, we use our previous experience to decide what to do next, to predict what will follow, or simply to characterize it as another instance of something with which we are familiar. What this means is that we can really only understand—and, hence, remember—situations we have been in before. Our memories are really little more than the sum of stories we can recall and apply. Part of knowing the right story to tell is having a lot of them.

THE EXPECTATION

The basic action of the human mind is to link one event to another. A baby who gets fed each time her mother appears in time links the sight of mother with food. We may come to relate some events in some abstract way, through the process of reasoning, but usually we connect events simply because they occurred close together in time and/or space. People learn to associate lightning and thunder long before they understand why they co-occur. After even a few exposures to phenomena, people begin to form expectations about them; after hearing, say, the screech of tires, most automatically expect to hear a metallic crash.

Forming expectations is a natural consequence of our search and retrieval process. The story we are reminded of by our current situation allows us to predict what will come next. To use a previous example, a person's experience with business promotions involving animal suits may lead them to predict that the man in the chicken suit will try to hand them a flyer. They do not have to spend much time reasoning about the situation in order to negotiate through it. Our ability to form expectations allows us to deal with the mind-boggling stream of information we receive about the world every moment. Rather than continuously monitoring every sensation, we can put part of our attention on "automatic," so to speak, and not have to worry about it. We all have an enormous number of expectations lurking in our heads, evidenced in our ability to quickly detect anomalies.

We form expectations at every level of abstraction, from visual perception to the beliefs of strangers. At a low level, we learn to link words and phrases, like "Merry" and "Christmas" and "How are..." and "you?" We sometimes finish sentences for people, certain we know what words are coming next. We form expectations at the conceptual level, associating the concepts of, for example, an object's speed and its potential force. Most jokes rely on people's tendency to have conceptual expectations, which are then twisted into a new meaning. And, of course, we are prone to connect events together, like seeing no taxis when it is raining, or blowing out the candles after singing "Happy Birthday."

In addition, we recognize patterns of regularity in the goals, plans, and behaviors of other people. We so often associate the goal of marriage with the goal of having children that we are surprised when people who achieve the former do not have the goal of the latter. We also form expectations with regard to the plans people use to achieve their goals. We know why someone would paint a room, because we know what outcome it will achieve (a better-looking room). Similarly, we do not have to ask someone why they own a car, or whether they have any means for doing laundry. We also expect other people to share our goals, and to use the same kinds of plans; if someone shifts jobs, we assume they do so for more money, increased challenge, or some such factor. Through repeated exposure, we come to anticipate other people behaving in certain ways, often summarized in a characterization like "untrustworthy," "friendly," or "long winded." These are our predictions of how they will act in future situations.

THE FAILURE

As a result of this process of linking events, we carry an extensive set of beliefs about future conditions of the world. When the world conforms to our expectations, we feel we understand it, and continue on. But what happens when our expectations fail? Learning begins. In a sense, when we succeed at something, we learn very little. Driving to work or eating in a restaurant are examples. But when we fail, our mind attends to the experience; we seek to explain our failures, and this forms the basis of our learning.

This follows from what we said before, that understanding is the process of applying past experience to the present one. When an expectation fails, it means that something did not match between the current situation and prior ones we have encountered: our conception of how the world should be, based on our past experience, does not correspond to how it actually is. Once we have detected a discrepancy, we choose to either integrate or ignore unmatched items. Often, we choose to ignore them. On the other hand, if we want to improve our performance, if we are pursuing a goal that is important to us, we will take the time to figure out what went wrong.

To understand this, consider a situation wherein someone bakes a cake (with nuts) that does not rise. They may choose to ignore the failure (attributing the problem to humidity or some twist of fate, perhaps) and, as a result, learn little from the experience. If they are interested in baking a cake with nuts in the future, however, they will expend the energy to figure out what failed. They may ask someone about it, or experiment on their own, divining causes and then confirming or refuting them. They might hypothesize that the nuts somehow retard rising, then gather evidence by noting that the nuts contain salt, and remembering that salt retards yeast growth. If they generate a plausible explanation, they index it in reference to the context in which the failure occurred, so the next time they use a recipe with nuts, they remember this episode, possibly adding extra baking powder or skipping the nuts to avoid the same problem. As people gain more experience doing something, their conception of it becomes laced with stories about failures and ways to avoid them. Experts, not surprisingly, can usually tell many stories about failures they have encountered, which they collect by participating in a given situation numerous times, running into obstacles, and then explaining and indexing them.

Failure works to change our conception of past events, as well. For example, say you observe an acquaintance leave a very low tip at a restaurant. This behavior may seem unusual to you at the time—was the service bad, or is the person cheap? If you later see him go all night without buying anyone drinks, you may put the experiences together and realize that he is a tightwad. This revelation then colors your original observation: you see his small tip as another instance of him being cheap. When a failure happens consistently, like anything else, we soon come to expect it.

Failure, then, drives the learning process. Without it, our sense of the world would be fairly immutable: we would form more and more expectations, each remaining forever active. Failures provide the spur for us to change our memory organization to better conform with the observed world. Of course, for this to work, we must form an expectation and, furthermore, detect a mismatch between it and our observation. When students move into an unfamiliar domain (such as taking their first physics class), they cannot begin to learn until they begin forming expectations and detecting failures. Because expectations lead to failures, which leads to learning and, hence, the forming of new expectations, the process is exponential: that may be why educators often speak of a "learning curve" involved in mastering a new area.

Search, expectation, and failure constitute the basic mechanisms by which human memory is modified, from visual images to abstract principles. When faced with a situation, people search for a similar incident in their past, and use it to form a prediction. Later, they compare that prediction to reality; when there is a mismatch, and they are interested in finding out why it occurred, they investigate it and so temper their expectations. This is how people reason about the world, and it is the way they naturally learn from it.

NATURAL LEARNING
LEARNING BY DOING

From everything we have said about how people make sense of the world, it is obvious that natural learning means "learning by doing." The way to learn how to do something is by doing it, which gives people access to the details they need to construct and index stories and, hence, to remember them. No amount of reading driver's manuals or watching films about driving will ever substitute for time behind the wheel of a moving car. If you want to learn how to sail a boat, build an airplane, or manage a business, you have to sail, build, or manage. Employers recognize this, knowing that workers become better at a position to the extent they gain experience in that position. The vast majority of professional development is "on-the-job" training, and want ads often stipulate that applicants have a certain number of years in the position (rather than, say, a degree in the field). Nearly everyone would agree that experience is the best teacher. What many fail to realize is that experience may well be the *only* teacher.

Why is learning by doing so much different than other types of learning, such as learning by being told? When you engage in an activity—and only when you engage in an activity—one of three things can happen:

- You reinforce previously linked actions.
- You fail to accomplish the goal.
- You accomplish the goal but cause an expectation to fail.

Each of these contribute to learning, in slightly different ways.

REINFORCE PREVIOUSLY LINKED ACTIONS. As you perform a task over and over, such as driving a car, juggling, or swinging a baseball bat, it becomes easier and more natural to do. We call this "getting a feel" for it. Through practice alone, a person so embeds the steps in memory that he can perform them automatically. In time, each step calls the next automatically. Typists and pianists whirl through complex sequences of finger moves much more quickly than they can consciously attend. In fact, people often have trouble retrieving a particular action in a sequence *except* in reference to the other actions in the sequence. Ask someone which leg they raise when releasing a bowling ball, for example, and they may need to act out bowling a frame in order to answer. Ask someone whether they set the temperature before or after they set the timer on their microwave, and they, too, may have to go through the steps in their mind. In both cases, people have stored a sequence of actions as a unit, so much so that they cannot treat one step except in the context of the others.

FAIL TO ACCOMPLISH THE INTENDED GOAL. In life, our best-laid plans often go astray. When they do, we figure out what went wrong so we can do better in the future. As a parent, you realize your children have to make mistakes on their own, in spite of the benefit of your ample experience. You just hope they do not kill themselves in the process. As we have stressed, learning is driven by failure. When a child tries to build a bridge and it collapses, he will try something different, and keep trying until he succeeds. He does this because he is interested in having a bridge. He learns what he needs to in order to achieve that goal. When a plan fails, we store the failure together with how to fix the plan, often in the form of a story. By building up our collection of stories, we remember how to avoid failures.

ACCOMPLISH THE GOAL BUT FAIL AN EXPECTATION. Often, our activities have unforeseen consequences. We apply herbicide to kill weeds, only to find it kills desirable plants as well. We discover that a certain dish is delicious, but makes a big mess in the kitchen. Becoming adept at something means knowing the consequences of our actions, both intended and unintended. When we next engage in the activity, we remember what to watch out for so we can guard against it. The difference between this outcome and the one above is more the scope of our learning, rather than the type. When a plan fails, we often make major changes to our conceptions of the situation, such as labeling the plan as worthless and abandoning it, or making major corrections to it. When we achieve a goal but fail some other expectation, in contrast, our learning is more an act of refinement and clarification. We may simply add an additional step to counteract a side effect (such as opening a window before cooking a smelly dish), or link two plans together to guard against a later consequence (such as sweeping the floor after we dust the furniture). As a matter of fact, it is this type of incremental knowledge that usually differentiates novices from experts in a field—"the devil is in the details," as they say. For instance, most people are generally aware of the outcome when someone is fired (e.g., the person is often bitter, and management has to hire someone new). Experienced managers can anticipate *all* of the possible ramifications of firing someone, including worried and demoralized co-workers, political manipulation by subordinates eager to fill the vacancy, increased workman's compensation obligations, and so on. In turn, good managers have contingency plans for dealing with each of these consequences, such as calling a meeting to reduce worry.

Learning by doing is the way we naturally learn. People are very good at learning on an as-needed basis, integrating knowledge as it pertains to accomplishing their goals. By repeatedly practicing an activity, people learn every step in the context of other steps. They become able to execute a complex routine with very little concentration. When we fail to achieve our goal, we figure out why and augment our memory accordingly. In time, our memories are a rich web of goals, plans, expectations, failures, and fixes.

The problem with learning by doing is that we lack a convenient means for describing the doing. Traditionally, pedagogical goals have enumerated facts and concepts (couched in terms like "Student has an awareness of X," or "Student knows X"), because it is relatively simple to do. It is difficult, on the other hand, to articulate the things students should be able to *do*. To properly shape our educational strategy, we must develop a vocabulary for our pedagogical goals.

HOW MEMORY IS CHANGED BY DOING

In *Scripts, Plans, Goals, and Understanding*[1] we established the concept of a "script," a way people organize knowledge so they can get around in the world. For instance, if Joe tells us about the time he ate sushi, we can assume that he was in a sushi bar, that he used chopsticks instead of a fork, and perhaps that he drank Japanese beer. We know this because we have a sushi-eating script. If we did not have this script, we would probably not be able to understand him very deeply.

Scripts enable people to successfully negotiate the situations they are in, often without much thought. People acquire scripts by learning them, usually by doing them over and over. We learn many of them as children, such as being taken to a restaurant many times and gradually developing a restaurant script. When an aspect of a script fails to come true (e.g., chopsticks at a sushi bar), we modify our scripts accordingly. Each time we engage in a script, we become better at executing it, at overcoming obstacles and applying it to other, less similar situations. Like stories, scripts are manipulated by the basic processes of search, expectation, and failure.

In *Scripts, Plans, Goals, and Understanding*, scripts are very large, rather immutable structures, such as going to a restaurant and taking a flight. However, such a construct does not describe how we are able to adapt to novel situations with relative ease, by applying what we learn in other situations. For example, someone who has paid for and used a ticket to get into a movie theater would have little trouble paying for and using a ticket to get into a ballgame. For people to be able to do this, their memory must be made up of individual scenes such as "buying a ticket" that could be flexibly assembled in memory by larger memory structures. In *Dynamic Memory*,[2] such scenes are called *MOPs*, for *Memory Organization Packets*. For example, we all probably have a buy-a-ticket MOP which we activate when we, go to a movie theater or ballpark; it can be divided into scenes such as "handing over money" and "putting the tickets away." Some MOPs remain fairly consistent across situations and contexts, while others are more idiosyncratic.

The important aspect about MOPs is that they can be extensively altered and reorganized, in light of new experience. When something unexpected happens, we locate the particular MOP that led to the failure and modify it. This then improves our performance in all situations that make use of the repaired MOP. For example, suppose you are paying at a restaurant and notice, when you get the credit card form, that they have changed the location where you are supposed to sign. Learning how to deal with the new form affects more than just your restaurant MOP, it affects each of the MOPs that make use of the "pay-by-credit-card MOP," such as booking flights and buying gifts.

In *Dynamic Memory*, scripts took on a smaller scope. The original version of a script might be dining in a restaurant, while the later version might be looking at a menu, putting ketchup on french fries, or figuring out the tip. People, however, continue to think of a script in the original sense. Hence, we have taken to calling this newer conception of the script "scriptlets," to emphasize the narrowness of their focus.

Scriptlets are central to any discussion of learning by doing. First, learning by doing causes scriptlets to be formed. When you sit down to figure out how to tape on your VCR for the first time, you slowly acquire the actions in which you must engage (put a tape in, rewind it, set the channel, set the timer, etc.). Over time, those actions become routine: you have acquired the scriptlet for taping on a VCR. When you say you know how to do something, like program your VCR, you are referring to the scriptlets you have acquired. Over time, people build up an extensive repertoire of scriptlets, for performing at their job (a business manager may have scriptlets for raising a person's productivity, analyzing sales' performance, and so on), dealing with their family, fixing their house, entertaining guests, and paying bills.

Often, we cannot easily describe our scriptlets, which generally contain an assortment of low-level actions we have practiced many times. As mentioned earlier, through repetition activities like swinging a baseball bat, playing a piano, and typing become automatic, to the point where the only way a person can think about a particular step is by tracing out the entire routine. This is true for purely mental scriptlets as well, such as a manager's scriptlet for firing someone.

Learning by doing not only helps us to form scriptlets, but to link different scriptlets together so we can reason about novel situations. Must of us have shopping scriptlets that are so generalized we can negotiate new stores with ease. To return to our previous example, VCRs resemble tape recorders in many ways: both have tapes to put in and record on, a set of buttons for controlling tape movement (play, rewind, fast-forward, stop, pause, and eject), a timer, and so forth. When people purchase their first VCR, they may apply the scriptlets they acquired for using a tape recorder. To succeed, they must modify portions of the tape-recording scriptlet, such as omitting the action "open the door of the tape chamber before inserting the tape" and adding the new action "set the channel using the remote control." But most of the tape-recording scriptlet works fine. If they buy a second VCR, they may have to generalize their taping-on-a-VCR scriptlet: perhaps the old VCR requires them to set the end time, but with the new one, the length of time to record. With each new VCR people use, their VCR-using scriptlets become more and more generalized, so much so in fact, that they can probably be used to operate videodisk players and the next generation of media.

To take another example, mid-level business people sometimes take new positions in different companies, even in different industries. They are able to settle into their new role surprisingly quickly (certainly in a fraction of the time they took to learn their original job), even though they might be selling a completely different product or service to a separate group of customers, work in an unfamiliar layout, and interact with a new set of co-workers. How do they adapt so effortlessly? It is because over the years they acquired an enormous number of scriptlets—writing business letters, giving orders to subordinates, making copies, analyzing markets, and so on. Through switching positions within a company, they have modified these scriptlets to be as broadly applicable as possible. They have learned, in other words, how to be a business person regardless of context. To the extent that similar positions require similar scriptlets, these people can perform.

Learning by doing changes our mind in a third way, too. It provides a structure by which we remember exceptional cases, such as expectation failures and unforeseen consequences. Many of our most memorable stories center on a failure that occurred while we were engaged in a scriptlet. "I was on my way to work...," we might start, thus letting the hearer know to which scriptlet we were referring, and then go on to tell about something out of the ordinary that happened. Sometimes stories are bound so tightly to a scriptlet that people can remember them only when they are engaged in a scriptlet. When people return to school after working for many years, they may suddenly remember a lot of long-forgotten stories, triggered by scriptlets related to attending school.

Learning by doing, then, causes us to form scriptlets, broaden their applicability, link them together, and remember them along with stories concerned with their deployment and use. Natural learning means learning scriptlets. But there is more to natural learning than simply doing. There is one critical element of learning we still need to consider.

GOALS DRIVE LEARNING

People do not learn automatically from every situation they are in. Much, if not most of their day is spent engaging in activities that do not cause any significant change in their memory. Why do people learn from some situations and not others?

The answer is simple: people learn when they want to learn. And they want to learn when doing so helps them perform some useful function. Failure drives learning, but a person who does not care whether he fails or not learns little. This raises a fundamental rule about learning by doing: it will not work if students do not want to learn. After all, asking questions, explaining anomalies, and correcting mistakes take effort. People need a good reason to expend that effort. When the reward is minimal (for instance, when our scriptlet for shopping fails because they moved the peanut butter to a different aisle), we do not spend much time trying to explain the change and, hence, to be enlightened by it. We may not even notice. It is just not worth the effort.

On the other hand, how people perform at their job *is* generally important to them, and so it is worth it for them to learn as much as possible. Professionally, people spend a lot of time recognizing and explaining their failures and, hence, building their repertoire of scriptlets. They do this because they are pursuing goals that are meaningful to them (such as increased salary, professionalism, etc.) and, in order to achieve those goals, they seek to avoid failures. Over time, they discard bad plans and automatize good ones, because the increased productivity is worth the effort.

There is another reason as well why goals are so essential to learning, apart from motivating us to expend mental effort. Goals play a role in the way we organize our memory. The vast majority of our stories reflect goals we have had. We remember how our plans screwed up, or cleverly succeeded. People do not learn how to avoid a failure except by recognizing the failure when they see it; when there are no goals, there are no failures. Given this, it should not be surprising that goals are the most important element in much of our indexing. Our scriptlets and stories are linked by our goals, which enables us to recall them quickly when a goal becomes active or threatened. This no doubts shapes our memory in a way beneficial to our survival.

The implications for education of our assertion that goals drive learning should be apparent. Learning in the absence of goals cannot modify memory in a meaningful way. If most of our experience is organized around our goals, it follows that we learn little when goals are not involved. Without goals there are no failures, and without failures there is no learning. Some people may argue that students in today's schools *do* have goals. While this may be true, many of those goals are quite artificial, such as getting good grades, not looking foolish, or pleasing the teacher. Such goals dictate the type of learning that will occur. Many teachers have complained that students forget much of what they learned soon after taking a test. The reason for this is obvious: they learned it only for the purpose of taking the test. It has no connection to life outside the classroom, so once the test has passed, so do the indices.

Memorization, especially, goes against the grain of natural learning, because it ignores the whole idea of functional knowledge, that is, knowledge we acquire for the purpose of doing useful things. When students learn something by rote, such as a list of vocabulary items for the SAT or GRE test, they are actually constructing very simple indices to it. They may form an association between a word and its synonym (such as "soporific" and "sleep-inducing"), grilling themselves with cue cards so that, in time, when they see one word, they automatically recall the other. Simply recalling a word's synonym, however, helps students in few contexts outside of taking the SAT or GRE test. It can be linked to few, if any, personal experiences. This is the reason so much of what we learn in school is quickly forgotten: when we do not richly index an item when we first store it, we have little way to remember it later.

Replace students' artificial goals with ones they already have, and their learning will change. A student's deeply held goal is likely to be richly connected in memory, and can thus serve as the basis for integrating new scriptlets and stories. A student interested in building an FM radio, for example, likely has a fair amount of knowledge of electronic components and how they fit together (perhaps by taking gadgets apart), of electricity (by using appliances), and the target condition (he wants music coming out of the speakers, and the ability to tune in to favorite stations). Thus, the student has a solid framework in which to integrate scriptlets involved in electronics.

NATURAL TEACHING

The discussion thus far has focused on how people naturally learn. We learn scriptlets and stories because they help us to achieve something tangible. We learn when we actually do things, not simply read about or view them. Activity gives us rich material for integrating the information into our existing knowledge and for recalling it later. Having said all this about learning, it raises the question, can scriptlets and exceptional cases be taught directly?

Many educators will answer yes, and point to the presumed dilemma sometimes called the "bootstrapping" problem: how can students learn by doing when they do not know how to do what they have to do to learn? Isn't a certain amount of learning-by-being-told necessary, at least in the beginning—we do not just put pilots in an airplane and tell them to fly, do we? There is really no dilemma here at all, however. Two-year-olds do not need to be given walking and talking lessons. They may have certain instinctual proclivities toward walking and talking, but they learned a great deal about both on their own, without formal school training. They learned because they wanted to move around and to communicate with others. They learned because they wanted to better participate in the world.

On the other hand, parents *have* successfully taught their children useful scriptlets and stories since humans began walking upright. When we make a mistake or detect an anomaly, it helps to have someone on hand to answer questions and tell us what we did wrong, and why. Obviously, some teaching works. The kind that does we call "natural teaching." Natural teaching puts students into situations where they learn by doing, allowing them to use the same techniques they used to acquire the bulk of their skills, such as experimentation, explanation, and story construction. Natural teaching provides instruction reactively, in response to student action.

Other methods of teaching—delivering facts and concepts to students apart from any functional context (such as making students memorize the Pythagorean theorem without showing them how it is useful), teaching each step of a procedure individually (such as teaching them vocabulary, then sentence diagramming, then paragraph construction), and even relying on lecture (where students passively listen for an hour)—conflict with the way students naturally learn. If a method does not allow students to participate in meaningful activity, in pursuit of their own goals, it does not tap into people's instinctual ability to profit from an experience. While some students may adapt successfully to it, most cannot.

THE GOAL-BASED SCENARIO

For natural teaching to be useful to educators, and to be replicable across different physical layouts and between participants, it needs a formal structure. We call one such structure "goal-based scenarios," or *GBSs*. A GBS is an educational environment which allows students to practice doing realistic activities in the pursuit of meaningful goals. A student plays some realistic role, such as a doctor on the Shuttle, a CEO, or a wildlife biologist on a game preserve, in order to accomplish a goal, or "Mission," that is consistent with that role. The student may be trying to hunt down the cause of a disease on the Shuttle, to increase a company's profits, or to protect a favorite animal. Everything in a GBS—the activities the student engages in, the setting, the feedback the student receives on his performance, and the like—is congruent with the Mission and the role. GBSs are a way to structure a learning-by-doing activity so that a student has control over it, while still learning useful scriptlets and stories annotated to them.

There is an enormous number of roles in the world, any of which might be incorporated into a GBS, and roles often embrace a variety of goals. Instructional designers thus have wide latitude in operationalizing a GBS. But, there are constraints. For a GBS to succeed, it must meet the four qualifications applicable to all natural teaching environments:

1. *Instruction should be one on one.* Since people make sense of current experience based on their past experience, and each person has a different set of experiences, it follows that teaching everyone the same thing, at the same pace, cannot work. Students learn naturally on an "as-needed" basis, when they stumble into a difficulty while doing something or raise a question they cannot answer. As a consequence, an instructor cannot assume that everyone will make the same mistakes, at the same time, but must instead wait for a particular person to make a particular mistake, and only then intervene. Instruction should be available at the time—and only at the time—a student needs it.

2. *Students should be able to explore in pursuit of their goals.* When something interests us, we actively investigate it, fiddling around with it, asking questions, and learning much. Most people learn how to use electronic mail, for example, by trying different commands and noting their effects. An educational environment should support trial and error. Moreover, since it is hard to determine what sub-topics a student may want or need to pursue, it should be able to entertain a variety of possible tangents. In short, an educational environment must offer a world sufficiently large for fruitful exploration. It is, of course, hard to provide a complex, compelling, exploratory environment for everyone. But it is necessary.

3. *Instruction should be delivered on demand, preferably as stories.* When engaging in a new activity, from remodeling to taxes, sometimes you get stuck. When you need help, you know it, and you ask for it. At that point, it is nice to have an expert nearby to address your specific concern. Good coaches and teachers know that stories are an especially effective way to instruct, often about overcoming mistakes the student is making. Stories stick in students' minds, because they can be richly indexed, and are plainly applicable to the immediate situation. If intelligence is the ability to tell the right story at the right time, students will benefit from intelligent teachers.

4. *Failure should drive instruction.* Failing to achieve what we expected drives our thinking processes: we sit up, take notice, and ultimately learn. Our memories serve to link together events, to raise expectations about them in the future, and, when those expectations fail, to figure out what went wrong. We do not have the time nor capacity to attend to and record every aspect of any situation, so we must choose which parts to integrate and which to ignore; a big part of expertise is knowing what to pay attention to. Since failure drives learning, it must also drive instruction. As we said, when we need help, we know it and ask for it. We know it by recognizing our failure and not knowing how to overcome it, such as when we cannot figure out why the glue is not sticking to the floor or whether we have to file a Schedule C. Effective teachers know how students can fail at an activity, they know why, and they know how to help.

THE GOAL-BASED SCENARIO AS A COURSE OF INSTRUCTION

There are obstacles to implementing a goal-based scenario in today's classroom. With what we have said about the way people learn, it is difficult to defend compartmentalizing knowledge into distinct disciplines like math, biology, and history, and dividing up activities into distinct periods over a fixed duration. That just is not how the world—apart from academia—operates, and it does not conform to natural learning. But courses, periods and semesters are a reality that must be addressed. Most students and teachers have grown used to a regular, punctuated schedule, and teachers are currently oriented toward particular disciplines.

A second constraint on implementing a GBS in the schools involves their educational charter: there are a great many things we would like all students to become skillful at, such as voting, making consumer purchases, and so on. How can we allow students to pursue their own goals and still learn what we wish? As every parent knows, it is sometimes impossible to harmonize what children want to do with what adults want them to do. Is not learning by doing a chance for students to do nothing at all?

A GBS can be structured to satisfy these constraints, that is, be a semester-long, hourly course that teaches a set of target skills that we, as a society, deem appropriate. It will take clever engineering, but it can be done. And in a way, it is not all that complicated: there are just two steps.

STEP ONE. Enumerate and refine the natural goals students already have, or can be induced to have, choose one or a few, then give them some structure and a tangible outcome. Children already have a lot of goals, from bashing trucks to designing clothes. They are fascinated by astronauts and firemen and other roles in the adult world. Many are intrigued by dinosaurs, dancing, sports, and the natural world. Tapping into these interests is like finding oil.

STEP TWO. Embed what you want students to be able to do into the GBS such that learning how to do it helps them achieve their goal. If a student wants what acquiring a skill will bring him, he is motivated to acquire it. Everyone wins: the student gets what he wants, and learns what we want. Of course, this is where the cleverness comes in. It is sometimes difficult to figure out how to teach certain things by doing. For instance, in Illinois a proficiency exam is given to all high school juniors on the Illinois Constitution: how can you require students to know facts like when Illinois became a state within the context of some plausible activity?

The difficulty we face integrating our goals and those of the students may be advantageous, however. There are more than a few sacred cows in our curricula, from the teaching of Latin to spelling drills, often taught simply because they have always been taught. Devising a GBS is a great exercise for rethinking what students must learn. It shifts us away from treating knowledge as a set of facts, concepts, and principles akin to a particular domain, to treating it as a process, intimately linked to activities we engage in. To embed knowledge of the Illinois Constitution in a GBS, we must first figure out what knowledge about the Illinois Constitution helps us to *do*. If we can answer this question, students should be learning about the Constitution while doing it. If, however, we cannot answer this question, we must seriously reconsider whether we should teach it at all. It does not seem unreasonable to require educators to justify what they teach by demonstrating how it helps students do something useful.

LEARNING BY COGNIZING

When trying to conceptualize the pedagogical goals of a course, sometimes we find that what we want students to do is not "doing" at all. Compare learning to tape on a VCR with learning to vote intelligently in an election. Taping on a VCR is relatively straightforward: each step (e.g., put tape in, set channel) is a physical action, easily observable, fairly distinct, and completed in a relatively fixed order. Choosing for whom to vote in an election, in contrast, seems very different. Though it does involve some physical action (e.g., go into the booth, punch holes by the candidate's name), that action is trivial compared to deciding which candidate to support. The task of selecting a candidate is not decomposable into a set of fixed steps: voters do not compare candidates' voting records, then their education and background, then their alliances, then make a decision, in that order. There are no hard and fast rules, just things to consider.

Above and beyond the doing, there is the thinking about the doing. A lot of the things we generally agree are important to teach fall into this category: purchasing wisely, making a smart business decision, analyzing a business plan, debugging code, writing a report, and so on. Such skills, which we call "cognitive scriptlets," focus on deriving new information not immediately present in a situation, and are composed of reasoning processes such as inferring, calculating, concluding, considering, and estimating. They are learned the same way we learned our scriptlets for cooking dinner, that is, by engaging in meaningful, goal-directed activity. For example, by making grocery lists, to-be-done lists, and such, we learn that items on the list need to be at the same level of specificity, that items can be sorted and ordered, and other aspects of the list-making scriptlet. Cognitive scriptlets are, like other scriptlets, tricky to decompose, because we automatize them just as we do, say, typing. For instance, when someone tells you they just bought a Weedeater for $20, you automatically infer that they have a lawn that needs trimming, that they now own a Weedeater, and that $20 is probably a deal. You infer these things so automatically that you probably are not even aware of it.

One class of cognitive scriptlets are those that relate to large, complex systems, like politics and economics. In the case of economics, millions of Americans own their own business. Each knows something about how the markets they participate in work, such as how to establish pricing, how raising prices will affect sales, who the major players are, what customers are looking for, and so on. Yet, few of these people have an intricate understanding of how the market as a whole operates: they cannot map out the demographics of all the major markets, establish the relationship between, say, interest rates and the price of commodities, compute the nation's money supply, nor even tell you how knowledge of the nation's money supply is useful. The knowledge that business owners have of the domain of economics is grounded in the context of operating their business within a larger system. Knowledge of a domain means being able to operate in that domain. Instruction that treats economics as a singular, knowable entity that can be mastered, as something that students can learn by being given a simulation of it as a whole and allowed to "play God" by tweaking parameters such as interest rates, foreign investment, and such, gives students a very different perspective than one that comes with being a small business owner. Someone with a "business mind" does not have a holistic sense of the overall economic system, but rather knows how to perform in roles within it.

Another class of cognitive scriptlets involves decision making. We all face a dizzying number of decisions in life, and we want to equip students to make informed decisions about their health, their purchases, the environment, political candidates, etc. Too often, though, decision making is taught in one of two ways, either telling students the "right" choice, or stressing the fact that there *is* a choice (as in preaching that "your vote counts"). Yet, decision making is an activity, just as skiing is: there are some useful heuristics or strategies associated with it (e.g., cost-benefit analysis), you get better at it with practice, and learning to do it means learning to avoid failures (such as not considering all the choices). Like skiing, then, there is only one natural way to teach decision making: have students do it. Knowledge for decision making means allowing students to make real, meaningful decisions. To profit from practice in decision making, students need to view the outcomes of various decisions. They must recognize how much information they must gather, and to recognize their options. Issues such as environmental protection and birth control become important when they connect with some pertinent decision in life, such as whether to allow an incinerator to be built nearby or how to use a condom.

A third class of cognitive scriptlets is often called "meta-cognitive strategies" or "higher-order thinking skills" (e.g., inferencing and analogizing). Since we cannot anticipate all of the situations students will encounter in their life, we must try to improve their ability to apply past experience to novel situations. Knowledge for life means thinking. We must get students to reason about and communicate their activities and, hence, to generalize their applicability. Thinking about the doing takes as much practice as the doing does. It is important to note that giving students practice thinking about their performance is not the same as teaching them thinking skills explicitly (such as in having them memorize Polya's problem-solving rules); thinking requires a context, a problem to investigate, a situation to assess. Thinking means tossing around facts in the pursuit of something desired.

SIX COURSE DESIGNS FOR NATURAL TEACHING

Having characterized how people naturally learn and the general principles by which a teaching method can augment the process, it is useful to explore illustrative examples of natural teaching. Below we describe six natural teaching designs, each highlighting a particular teaching strategy and embodied in a working piece of software. No doubt, each of the systems we describe employ multiple teaching strategies;

nevertheless, each system offers a salient view of one in particular. The six teaching strategies are as follows:

- Role Playing
- Accomplishing a Task
- Incidental Learning
- Learning by Cognizing
- Exploratory Learning
- Providing Help

NATURAL TEACHING DESIGN #1: ROLE PLAYING

In a role-playing goal-based scenario, the student assumes some realistic role. For new employees, it would probably be the job for which they have been hired. For younger students, it should be a role that has meaning, such as being a voter or the CEO of a business. Everything the student does in the GBS is done in conjunction with the role: the student's goals, activities, etc. are all congruent with it. If the role is associated with a complex system (such as a market), the student is exposed to that system in the context of the role.

The particular natural teaching strategy engendered in a role-playing GBS is failure-driven instruction. As the student performs in a role, he makes mistakes. This, in turn, drives instruction. The pedagogical focus is on enabling the student to perform in the role without making a mistake, to adopt the scriptlets that pertain to achievement of the Mission. For instance, a student playing a CEO might underprice his product (and thus not make a profit), overestimate demand, or spend too little on marketing; when the student makes one of these mistakes, the GBS structure must ensure that the student profits from the mistake (e.g., by making sale figures available or modeling techniques to forecast demand), so he can avoid the same mistake in the future. With practice, students come to perform the role flawlessly. This is the model by which people are taught many things, such as how to swing a golf club: the student swings and the pro responds when he recognizes a problem with which he can help.

Dustin is a language training system developed for non-English speaking employees of Andersen Consulting who are about to come to the United States for training at Andersen's corporate educational center in St. Charles, Illinois. Andersen Consulting needed a way to smooth an employee's integration into the St. Charles environment. Students must be acquainted enough with the Center, and proficient enough at English, that they are able to perform day-to-day tasks such as getting their mail and participating in the classroom. Ideally, this could be solved by giving students some time to become comfortable around the Center before beginning classes, but the cost of this solution is prohibitive.

Dustin prepares students for the St. Charles environment by having them play the role they will later fill, that of newcomer to the St. Charles center. *Dustin* is a multimedia simulation of St. Charles. In it, the student has goals such as getting to the Center, checking into the hotel, and finding the classroom he is to train within. To achieve those goals, he must participate in various scenes (such as the customs line, the check-in desk, and his room), and interact with people (such as the customs agent, limousine driver, and receptionist). For instance, at one point in the program the student is in the front lobby of the hotel, wanting to check in. The screen shows the lobby. The woman behind the counter greets him, as she would in real life, except here the student sees a video of it. The student types his response in English, perhaps giving his name or asking whether he found the right place. The woman responds appropriately (looking up his name, answering his question, or whatever), and the dialogue continues. If the student succeeds in getting through the task, the system sends him off to do other things (such as going to his room and meeting his roommate). If the student fails, the system intervenes.

In *Dustin*, the student acquires scriptlets relevant to functioning at Andersen's training center, by encountering failures while performing tasks. For instance, the student may not know how to take his clothes to a cleaners (or *that* he can take them to the cleaners). The student is then in a position to learn something, and *Dustin* can intervene in several ways. First, it can model the task by showing a video of someone executing it. It can break down the task, describing the various steps. It can provide a transcript of the student's performance, so the student can figure out what went wrong. It can also provide translations of the conversation in the student's native tongue. Most importantly, it enables the student to practice, as many times as is needed to get it right. *Dustin* can do these things because its instruction is indexed around the goals newcomers are likely to have, the plans they undertake to achieve them, and the failures that could result.

Obviously, *Dustin* dramatically departs from traditional methods for teaching a foreign language. It does not supply grammars, common phrases, nor abstract cultural information for students to memorize. Its instruction is always supplied within a realistic setting, so students index the information with regard to situations they will actually experience when they arrive. It does not try to address generic problems that students might potentially have; instead, it reacts to what problems they *do* have and, thus, delivers instruction from which students actually benefit. Perhaps most importantly, it offers a sufficiently realistic world for students to explore, a landscape rich enough to give them a chance to fail.

NATURAL TEACHING DESIGN #2: ACCOMPLISHING A TASK

Everybody wants to achieve things. Get a date. Build a house. Run a business. A student pursuing a goal that is important will gladly learn how to succeed. *Goal-directed learning* centers around a tangible, physical goal like building a tree house or designing a dress. With the goal plainly in sight, students can focus their energies on achieving it. They can continuously monitor their progress and, thus, detect failure. They work until they are done, and they know when they are done. There is a pride that comes with accomplishment. All this can be profitably employed by education.

The challenge in developing a goal-directed learning environment is to find a goal that students are interested in. It must be complex enough to be challenging, realistic enough so students can relate to it, and unambiguous enough that students can make fairly direct progress. Everything in such a GBS—the instruction, the activities, the stories—revolves around achieving the goal, so it follows that the goal must be well thought out. Tasks involving designing and building perhaps capture the essence of goal-directed learning: students learn scriptlets and stories in the context of producing an object that meets some functional criteria. The student, meanwhile, is happy to own the object, and immediately recognizes the viability of the skills that were learned.

Broadcast News is designed to teach high school students contemporary history by having them compose a news show. The students' goal is to produce a coherent segment, doing the announcing themselves via a camera connected to the computer. They are supplied with news feeds, raw footage, interviews, accounts, and so on; in the course of editing videos, selecting cuts, and writing text, students acquire knowledge about the events they are covering, and the difference in people's interpretations of them. They learn this, though their focus is entirely on television news production. To succeed, they must be able to put together different fragments into a polished presentation, which means comprehending the meaning and significance of the news they report. For instance, if they are editing a video clip that makes a reference to Prime Minister Gandhi, they may need to find out what "Prime Minister" means, or to what extent they must explain the term to their viewers. By figuring this out, they learn the meaning of the concept almost without thinking about it.

Note that in *Broadcast News*, the task (composing a news segment) is plausibly related to its pedagogical goals (understanding contemporary history). TV news is the context by which we get most of our information about the world, and only by actually participating in news production can people get a sense of the selective, interpretative nature of journalism. Students come to comprehend history in the context of television news, but this directly pertains to how most people comprehend it. In *Broadcast News*, as in life, history is literally what we make of it.

NATURAL TEACHING DESIGN #3: INCIDENTAL LEARNING

We learn a great many things in the course of our lives that are not directly involved in the pursuit of our goals. A corporate lawyer, to take one case, often learns a lot about corporate politics serendipitously, in the course of his affairs. Carpenters know a lot of geometry. People readily pick up things in such a way, usually without trying, perhaps without even realizing it.

Such "incidental learning," in fact, may be a useful way to teach many things for which a direct role or context cannot be devised. Majors in ecology, for instance, apply simple differential equations in the course of building population models; when it comes to the math, they learn it, though they may not know how to apply it to, say, calculating the displacement of a falling body. If skill at integral calculus is deemed important, students might be taught it in the context of an ecology GBS. Most real-world activities similarly offer a context for learning a variety of scriptlets and stories.

However, for incidental learningto be an effective teaching paradigm, there has to be a plausible connection between the task and the target domain. Since the task is the context by which students integrate new information, if it isn't legitimate, problems ensue. Some educational software, for example, employ a contrived task as a veil over subject matter, such as shooting down verbs or landing a lunar module by solving equations. This will not work, for two reasons. First, students are likely to see through the charade, and be turned off. Second, because the task provides the basis for constructing indices, students engaged in a contrived task are not likely to index the knowledge they acquire in a meaningful way, that is, in a way that makes the knowledge applicable to other tasks and goals.

To support an incidental learning environment, both the task and the tangents must somehow be embodied in the architecture. When a student shows interest in something tangential to the current task, good teachers recognize the opportunity and seize it, provided they know how the tangent fits into the larger context, have the time, and are qualified enough to discuss it. Computer-based environments must likewise be able to seize such opportunities; to do so, they must possess information about the task and contain numerous stories relating to it and the context of which it is a part.

In *Road Trip*, grade school students learn geography in order to travel around the country by car. The screen displays a giant map of the United States. The map is interlaced with interstates, likely to be the anchoring point for most people's experience with the United States. Students can also access blow-up maps of individual states and regions. The student's car is depicted on the map, starting at Chicago. Students must look over the map and choose destinations to which to drive, then plot out their course, using a mouse to click on the destinations they seek and the interstates they want to use. When they are ready to travel, the car moves along the selected route and a red line traces their trip, while a video in the corner shows moving scenery from a car window.

Upon reaching a city, *Road Trip* users can choose to watch video clips of landmarks, events, and people that give a region its character, such as famous buildings, sports teams, foods, and attractions. In New York, for instance, a student can watch King Kong carry Fae Rae up the Empire State Building. In Pennsylvania they can learn about the Amish by watching scenes from "Witness." Ideally, the important functions, salient characteristics, and general flavor of each city and region can be captured on video and made available, so a student can better relate to and remember the local color of different areas of the United States.

Students like to travel, and are motivated to get access to the regional videos. Using *Road Trip*, they learn how to read a map, how to figure out directions, and how to plan routes. They learn the major metropolitan areas, where they are located, and their position relative to each other. And they learn this in the context of how they will likely experience geography later in life, in a car figuring out directions. The scriptlets they acquire for planning trips and recognizing important regional distinctions have an immediate relevance to real-world activity. Students, in *Road Trip*, are interested in getting places, and the task is structured such that reading a map, plotting routes, and the like directly and visibly contribute to it. Students do not have to ask why they should learn, say, that Pittsburgh is east of Chicago; they recognize that they have to know this if they want to drive to Pittsburgh.

NATURAL TEACHING DESIGN #4: LEARNING BY COGNIZING

Sometimes, we want students to become generally knowledgeable in an area, that is, to be able to think intelligently about an issue, like AIDS or Operation Desert Storm. The mistake we too often make is to try to teach it directly, such as distributing a pamphlet of "Do's" and "Don'ts." With everything we have said about natural learning, we know this will not work. We must instead devise situations in which students *want* to think about issues in a meaningful way.

Knowledge of a domain means operating within that domain. An educational environment that allows students to learn by cognizing must let students reason about an issue, and apply that reasoning to a real problem. While every environment should require students to reason about a domain, some will focus particularly on activities such as forming hypotheses and testing them.

Often, thinking intelligently about an issue means making the best-possible decision, that is, the one you gain most from. As citizens, family members, workers, and people on the street, we are all faced with countless decisions. Who to vote for. What to buy. Which plan to follow to get a better job. What classes to take. Many of these decisions revolve around prioritizing goals, such as sex and religious doctrine. Some involve choosing from competing plans, such as going to college or straight to work. In all cases, though, decision making involves considerations to make, sift through, and weigh. What are the consequences of the plan? What are the alternatives? What should I look out for? Which costs less? What is more important to me? What other effects might I anticipate?

In order to teach museum visitors about Sickle Cell disease, *Sickle Cell Counselor* places visitors in the role of genetic counselor. The visitor advises couples about the difficult decision of whether to have children when there is a risk of passing on the disease. A session begins with a video of a couple, discussing themselves and their hope for a child. To carry out their role as counselor, a visitor engages in four activities: taking blood samples from the couple and performing blood tests, computing genetic probabilities, asking experts, and advising clients. There is no mandated order to the activities; for instance, visitors can advise the client whenever they feel enough information has been gathered.

The system is complex and realistic enough to engage museum visitors for a considerable time. It stores videos of various experts, such as a pediatrician, a lab technician, and a genetic researcher, discussing various aspects of the situation from their own perspective. It contains simulated lab instruments with ample output, and graphic representation in the Punnett Square. Because it is realistic, visitors can navigate its screens fairly intuitively. *Sickle Cell Counselor* is structured around a decision the student is asked to make, that of deciding whether or not it is a good idea for the couple to have children. Students do not passively absorb facts about the disease; they must attempt to apply their understanding, in an authentic way. After all, the major reason we want people to know about Sickle Cell disease is so they can make an informed decision in light of it, and/or inform others facing such a decision. *Sickle Cell Counselor* takes a subject that is normally treated in an informational fashion, and transforms it into concrete activity. It does not discuss the effects of the disease abstractly; instead, it humanizes the issue, as the visitor sees how the disease may affect an actual couple. Rather than simply present facts about the disease, it gives users a forum in which to test out their ideas and reach conclusions. It offers an integrated context in which to learn about genetic testing, reproduction, etc. It plausibly requires students to consider all of the issues before they can make a valid decision, which entails understanding those issues in a deep, functional way.

NATURAL TEACHING DESIGN #5: EXPLORATORY LEARNING

Often, students are already motivated to learn about a domain, and know what they wish to learn more about. This is especially true in business. Trainees are already familiar with the position they are going to fill, and often have questions about it even before they begin training. The role already exists, so it does not need to be simulated; the task of education is to be able to answer students' questions, pointing out things to watch out for. Many educational software packages, such as the multitude of hypermedia stacks available, rely on an exploratory learning design. As such, they must assume that students are already motivated to browse, in other words, that students possess goals that they can satisfy with the information contained within the stack. Too often, however, that assumption is not warranted. If a student is not pursuing a goal, the student has no reason to think very deeply about the content and, hence, to learn. The student has no foundation on which to construct the knowledge thus obtained. Moreover, nothing guides the student in his/her search.

If the student already has a purpose in mind, on the other hand, the task of the teacher becomes one of responding to a student's questions, which means anticipating which ones might be asked, and having an answer. This takes knowledge of the student's role, and of failures the student might encounter. For this capability to be emulated in a computer-based environment, it must be indexed in an intuitive way, such that the student can easily pursue lines of reasoning.

TransAsk is designed to teach students about military transportation planning. It was built for the United States Transportation Command (TRANSCOM), a joint military command responsible for planning, coordinating, and scheduling military transportation. *TransAsk* is organized around a model of the job roles in the TRANSCOM crisis action team, the tasks carried out by people in those roles, and the problems typically encountered in successfully accomplishing them. Students using *TransAsk* have a particular role within TRANSCOM, have carried out tasks in that role, and have run into problems.

TransAsk contains video clips of experts telling stories about specific problems, issues, and consequences related to various tasks. For instance, a lieutenant colonel tells a story about delivering turkeys to Desert Storm on Thanksgiving; another talks about trying to arrange people to work on Christmas. These video stories are organized around the model of conversation described at the beginning of this piece. It supposes that when a student hears a story, he/she makes sense of it by recalling experiences that relate to it. The story may answer a question previously raised, it may place another in context, and so on.

With cleverly indexed video clips, *TransAsk* can tell stories that the student wants and needs to hear. This gives the student a way to integrate them with previously gained experiences. *TransAsk* features eight links between stories, representing typical follow-up questions. For example, after a student hears the "Working on Christmas" story, he may want to know the *Context* in which the story takes place; the system can respond with a story about who has final control in transportation planning. The student may want to hear Analogies to the story, such as the story about delivering turkeys on Thanksgiving. The student may wish to learn about the *Causes* of the events in the story, or their *Results*. Perhaps the student wants to hear more *Specifics*, such as a story about the amount of control TRANSCOM leadership exercises over day-to-day operations. The remaining links are *Alternatives* (other plans/options to achieve the same outcome), *Warnings* (things to watch out for in conjunction with the story), and *Opportunities* (advantageous situations raised by the events in the story).

With so many potential stories to choose from, a student might become overwhelmed if they are not presented in a comprehensible way. The eight links are always on the same place on the screen, radially extending from the central, current story. A follow-up story to the current one is represented as an index card with a brief summary. If there is more than one story for a given link (i.e., if there are two stories about *Results* of delivering turkeys on Thanksgiving), they are stacked. When a student sees a story that looks interesting, he can access it by double-clicking on its index card, causing the system to present the chosen video. That story then moves to the center of the screen, and follow-up stories are retrieved for it and displayed on the screen adjacent to the appropriate link. Each story thus allows students to follow any or all of those links, so the student can pursue topics deeply (such as tracing the causal chain that led to the current story) or broadly (such as hearing about the *Causes*, *Consequences*, and *Specifics* of a certain situation).

TransAsk does not actually engage the student in a task or simulate the role. It presumes that students already have experience with both. It instead serves more passively, giving students the freedom to pursue stories that interest them as deeply as they wish. Even so, it must model the role, failures, and questions of the student, so it can deliver the right story at the right time. It must logically present its videos so the student can find what he needs, and not get overwhelmed by the choices or lost in the stack.

NATURAL TEACHING DESIGN #6: PROVIDING HELP

When you encounter problems doing your taxes, remodeling your house, or investing some capital, you need to turn to an expert in the domain. You tell the expert your problem, and receive advice or other relevant information. This is the way most adults learn things in new areas, and it is very intuitive.

For an educational system to offer such help when needed, it must be organized around the tasks that students engage in, and hinged to the problems they are likely to encounter. It must relate instruction to the situation in which the problem occurred. And it must allow students to ask questions. If we want to provide help to people filling out their tax forms, we should be able to tell them that they should fill out Schedule B if their savings interest or dividends are above a certain amount, that using the standard deduction may lower their tax obligation, and so on. We have to be ready to answer if they ask about certain deductions or whether there are other forms they need to fill out. To be able to do this, we must know how to fill out tax forms correctly.

Arthur Andersen provides service in over 100 tax specialty areas for its worldwide clients. While most Andersen professionals specialize in one particular area (e.g., real estate), their clients often need consultation in multiple areas (e.g., purchasing a warehouse in another state requires knowledge of real estate, capital expenditure, the other state's tax code, etc.). *Taxops* is designed to teach Andersen professionals to think broadly for their clients and be able to recognize potential opportunities for additional consulting.

The student describes his client or potential client to the system by answering a number of concrete questions (nature of its business, its structure, and so on). *Taxops* then performs an analysis to spot opportunities. It may say "I have found a really good real estate opportunity, and pretty good opportunities for state and local tax, and personal financial planning," then offer to show video clips of experts talking about those opportunities. After being presented with a story, the student can choose to pursue it in greater depth.

Taxops assumes that its students have already had experience advising clients. Andersen professionals often encounter situations that extend beyond their area of expertise, so the problem is familiar to them. This reduces the burden on the system of addressing student's concerns, because its designers already have a pretty good idea what those concerns are. Hence, information is presented in a form of immediate use to the student.

Naturally, it is hard to get experts to give you every piece of advice that might possibly be relevant. You cannot anticipate every situation and, even if you could, no system could easily navigate such complex terrain. *Taxops* is limited by the stories it has to tell. If no expert addresses a concern, *Taxops* does not treat it. But the knowledge it possesses is maximally useful to the students, because it is tied to their needs.

NATURAL LEARNING AND NATURAL TEACHING

This account of human memory is fairly simple, yet makes plain how we should be schooling our youth. Human memory absorbs stories, by integrating them with past experiences and stories. It recalls them when circumstances warrant and uses them to comprehend the situation and predict the future. It is goal directed, with its contents organized around goals and plans for achieving them. It forms expectations, then revises them as new information comes to light. Human memory provides the next step in doing, executing scriptlets embracing complex activities such as typing and accounting. Each step calls the next without our conscious involvement. It also provides the next step in thinking, allowing us to quickly weigh situations, make decisions, and synthesize knowledge.

These are the basic processes by which we move about the world, and they are the processes by which we learn about it. Natural learning means learning by doing in the pursuit of goals, hearing stories and telling them, acquiring scriptlets and exceptional cases, and generalizing them to be more widely applicable and less flawed. Motivation and direct instruction simply are not relevant for two-year-olds learning to walk and talk. Their natural learning mechanisms work. There is no reason for it to be any different in our schools.

To be effective, teaching must complement natural learning. Natural teaching, therefore, tells stories relevant to students' concerns, when they are willing to hear them and can profit from them. It induces goals, real ones, in the student, so the student's behavior becomes goal-directed. It raises students' expectations, then causes them to fail. The failure happens in a natural way, so students can richly index the experience. Natural teaching supports situations in which practice in doing is desired. If students are interested in the goal and realize that getting better at something will help them to achieve it, they will practice. Through practice, they routinize activities until they can do them without thought. Natural teaching helps students broaden their skills by supplying opportunities to apply them in new situations. Natural teaching supports situations in which practice in thinking is desired, too. It enables students to develop strategies for reasoning effectively about a subject of interest to them, and apply that reasoning in some useful way.

The a question is not whether natural teaching is desirable, because natural teaching is the only way to support how we ordinarily learn about the world. Rather, the question is how natural teaching might be supported in the classroom. It is hard to imagine learning by doing in a room full of twenty-odd students, a book per person, and a single teacher alone, and there are not enough experts to provide individual counseling every time a student can benefit from it. The logistics of creating an authentic environment and rich experience for everyone are complex.

Computers can help. They can provide complex simulations, multimedia presentations, and tools for reasoning and calculation. They can store and distribute students' work. They can be remotely linked, to broaden participation. They can replicate successful environments in other schools and classes, preserving effective techniques. They can distribute expertise in a captivating way across great distances. They can, of course, be wrongly built, oversold, and misused. But designed correctly, using all the cleverness and expertise that instructional designers can muster, they can make natural teaching a reality.

Most of all, computers give us a fresh rationale for critically examining the educational enterprise. Computerized instruction calls for a different way of looking at the world. Neatly divided subjects, reliance on facts, and mastery learning must all be reorganized to fit the constraints of the technology, and thus reevaluated. Computers shift the focus of education from inert knowledge to knowledge of doing. Students become active and participatory, learning how to perform in the world. If for no other reason than this, computers are invaluable to natural teaching.

REFERENCES

1. Schank, R. C., and R. Abelson. *Scripts, Plans, Goals, and Understanding.* Hillsdale, NJ: Erlbaum Assoc., 1977.
2. Schank, R. C. *Dynamic Memory.* Cambridge, MA: Cambridge University Press, 1982.
3. Schank, R. C. *Tell Me A Story.* New York: Charles Scribner's Sons, 1990.

Stevan Harnad
Department of Psychology, University of Southampton, Highfield, Southampton, SO17 1BJ
UNITED KINGDOM; e-mail: harnad@ecs.soton.ac.uk

Does Mind Piggyback
on Robotic and Symbolic Capacity?

Cognitive science is a form of "reverse engineering" (as Dennett has dubbed it). We are trying to explain the mind by building (or explaining the functional principles of) systems that have minds. A "Turing" hierarchy of empirical constraints can be applied to this task, from T1, toy models that capture only an arbitrary fragment of our performance capacity, to T2, the standard "pen-pal" Turing Test (total symbolic capacity), to T3, the Total Turing Test (total symbolic plus robotic capacity), to T4 (T3 plus internal [neuromolecular] indistinguishability). All scientific theories are underdetermined by data. What is the right level of empirical constraint for cognitive theory? I will argue that T2 is underconstrained (because of the Symbol Grounding Problem and Searle's Chinese Room Argument) and that T4 is overconstrained (because we don't know which neural data, if any, are relevant). T3 is the level at which we solve the "other minds" problem in everyday life, the one at which evolution operates (the Blind Watchmaker is no mind reader either), and the one at which symbol systems can be grounded in the robotic capacity to name and manipulate the objects that their symbols are about. I will illustrate this with a toy model for an important component of T3—categorization—using neural nets that learn category invariance by "warping" similarity space the way it is warped in human categorical perception: within-category similarities are amplified and between-category similarities are attenuated. This analog "shape" constraint is the grounding inherited by the arbitrarily shaped symbol that names the category and by all the symbol combinations that it enters into. No matter how tightly one constrains any such model, however, it will always be more underdetermined than normal scientific and engineering theory. This will remain the ineliminable legacy of the mind/body problem.

Those who attended the conference and those who read the published volume of papers arising from it will be struck by the radical shifts in focus and content among the various categories of contribution. Pat Goldman-Rakic discusses internal representation in the brains of animals and Larry Squire discusses the brain basis of human memory. Others present data about human behavior, others about computational models, and still others about general classes of physical systems that might share the relevant properties of these three domains—brain, behavior, and computation—plus, one hopes, a further property as well, namely, conscious experience: this is the property such that, as our brains do whatever they do, as our behavior is generated, as whatever gets computed gets computed, there's somebody home in there, experiencing experiences during most of the time the rest of it is all happening.

It is the status of this last property that I am going to discuss first. Traditionally, this topic is the purview of the philosopher, particularly in the form of the so-called "mind/body" problem. But these days I find that philosophers, especially those who have become very closely associated with cognitive science and its actual practice, seem to be more dedicated to minimizing this problem (or even declaring it solved or nonexistent) than to giving it its full due, with all the perplexity and dissatisfaction that this inevitably leads to. So although I am not a philosopher, I feel it is my duty to arouse in you some of this perplexity and dissatisfaction—if only to have it assuaged by the true philosophers who will also be addressing you.

THE MIND/BODY PROBLEM

So here it is: The mind/body problem is a conceptual difficulty we all have with squaring the mental with the physical.[10,18] There is no problem in understanding how or why a system could be generating, say, pain behavior—withdrawal of a body part when it is injured, disinclination to use it while it is recovering, avoidance of future situations that resemble the original cause of the injury, and so on. We have no trouble equating all of this with the structure and function in some sort of organ, like the heart, but one whose function is to respond adaptively to injury. Not only is the structural/functional story of pain no conceptual problem, but neither is its evolutionary aspect: It is clear how and why it would be advantageous to an organism to have a nociceptive system like ours. The problem, however, is that while all those adaptive functions of pain are being "enacted," so to speak, we also happen to *feel* something; there's something it's *like* to be in pain. Pain is not just an adaptive structural/functional state, it's also an *experience*. It has a qualitative character (philosophers call such things "qualia") and we all know exactly what that experience (or any experience) is like (what it is like to have qualia); but, until we become committed to some philosophical theory (or preoccupied with some structural/functional model), we cannot for the life of us (if we admit it) see how any structural/functional explanation explains the *experience*, how it explains the presence of the qualia.[32,33]

Philosophers are groaning at this point about what they take to be an old canard that I am resurrecting to do its familiar, old, counterproductive mischief to us all over again. Let me quickly anticipate my punchline, to set them at ease, and then let me cast the problem in a thoroughly modern form, to show that this is not *quite* the same old profitless, irrelevant cavil that it has always been in the history of human inquiry into the mind. My punchline is actually a rather benign methodological one: Abstain from interpreting cognitive models mentalistically until they are empirically complete. Then you can safely call whichever you wish of their inner structures, functions, or states "qualia," buttressed by your favorite philosophical theory of why it is perfectly okay to do so, and how nothing is really left out by this, even if you may think, prephilosophically, that something is still missing. It is okay because the *empirical* story will be complete, so the mentalistic hermeneutics can no longer do any harm (and are probably correct). If you do it any earlier, however, you are in danger of *over-interpreting* a mindless toy model with a limited empirical scope, instead of widening its empirical scope, which is your only real hope of capturing the mind.

That is the methodological point, and I do not think anyone can quarrel with it in this form, but we will soon take a closer look at it and you will see that it has plenty of points on which there is room for substantive disagreement, with corresponding implications for what kind of empirical direction your research ought to take.

THE HIERARCHY OF TURING TESTS

In the 1950s, the logician Alan Turing, one of the fathers of modern computing, proposed a simple though since much misinterpreted "test" for whether or not a machine has a mind[37,38]: In the original version, the "Turing Test" went like this:

You're at a party, and the game being played is that a woman and a man leave the room and you can send notes back and forth to each of them, discussing whatever you like; the object is to guess which is the man and which is the woman (that is why they are out of the room: so you cannot tell by looking). They try to fool you, of course—and many people have gotten fixated on this aspect of the game (the attempted trickery); but in fact trickery has nothing to do with Turing's point, as you will see. What he suggested was that it is clear how we could go on and on, sending out different pairs of candidates, passing notes to and from each, and guessing, sometimes correctly, sometimes not, which is the man and which is the woman. But suppose that, unbeknownst to us, at some point in the game, a machine were substituted for one or the other of the candidates (it clearly does not matter which); and suppose that the game went on and on like that, notes continuing to be exchanged in both directions, hypotheses about which candidate is male and which female continuing to be generated, sometimes correctly, sometimes incorrectly, as before, but what *never* occurs to anyone is that the candidate may be neither male nor female, indeed, not a person at all, but a mindless machine.

Let's lay to rest one misconstrual of this game right away. It is unfortunate that Turing chose a party setting, because it makes it all seem too brief, brief enough so it might be easy to be fooled. So before I state the intuition to which Turing wanted to alert us all, let me reformulate the Turing Test not as a brief party game but as a lifelong pen-pal correspondence. This is fully in keeping with the point Turing was trying to make, but it sets aside trivial tricks that could depend on the briefness of the test period, or the gullibility (or even the drunkenness) of the testers on that particular night.[21]

Now the conclusion Turing wished to have us draw from his test: He wanted to show us that if there had never been anything in our correspondence that made us suspect that our pen-pal might not be a person, if his pen-pal performance was *indistinguishable* (this property has now come to be called "Turing Indistinguishability") from that of a real person, and, as I have added for clarity, indistinguishable across a lifetime, then, Turing wished to suggest to us, if we were ever informed that our pen-pal was indeed a machine rather than a human being, we would really have no nonarbitrary reasons for changing our minds about what we had, across a lifetime of correspondence, quite naturally inferred and countlessly confirmed on the basis of precisely the same empirical evidence on which we would judge a human pen-pal. We would not, in other words, have any nonarbitrary reason for changing our minds about the fact that our correspondent had a mind just because we found out he was a machine.

Be realistic. Think yourself into a life-long correspondence with a pen-pal whom you have never seen but who, after all these years, you feel you know as well as you know anyone. You are told one day by an informant that your pen-pal is and always has been a machine, located in Los Alamos, rather than the aging playboy in Melbourne you had known most of your life. Admit that your immediate intuitive reaction would not be "Damn, I've been fooled by a mindless machine all these years!" but rather an urge to write him one last letter saying "But how could you have deceived me like this all these years, after everything we've been through together?"

That is the point of the Turing Test, and it is not a point about operational definitions or anything like that. It is a point about us, about what we can and cannot know about one another, and hence about what our ordinary, everyday judgments about other minds are really based on. These judgments are not based on anything we know about either the biology or the neurobiology of mind—because we happen to know next to nothing about that today, and even what we know today was not known to our ancestors, who nevertheless managed to attribute minds to one another without any help from biological data: In other words, they did it, and we continue to do it, on the basis of Turing Testing alone: If it is totally indistinguishable from a person with a mind, well then, it has a mind. No facts about brain function inform that judgment, and they never have.

But does Turing Testing really exhaust the totality of the facts on which judgments about mentality are based? Not quite. Another unfortunate feature of Turing's original party game was the necessity of the out-of-sight constraint, so that you would not be cued by the *appearance* of the candidate—in the first instance, so you could not *see* whether it was a man or a woman, and in the second instance, so you would not be prejudiced by the fact that it looked like a machine. But in this era of gender-role metamorphosis, sex-change operations, and loveable cinematic robots and extraterrestrials, I think the "appearance" variable can be safely let out of the closet (as it could have been all along, given a sufficiently convincing candidate): My quite natural generalization of Turing's original Turing Test (henceforth T2) is the "Total Turing Test" (T3), which calls for Turing-Indistinguishability not only in the candidate's pen-pal capacities, in other words, its *symbolic* capacities, but also in its robotic capacities: its interactions with the objects, events, and states of affairs in the world that its symbolic communications are interpretable as being *about*. These capacities too must now be totally indistinguishable from our own.[13]

Unlike the still further requirement of neurobiological indistinguishability (which I will now call T4), T3, robotic indistinguishability (which of course includes T2, symbolic indistinguishability, as a subset) is *not* an arbitrary constraint that we never draw upon in ordinary life. The symbolic world is very powerful and evocative, but we would certainly become suspicious of a pen-pal if he could never say anything at all about objects that we had enclosed with our letters (e.g., occasional photographs of our aging selves across the years, but in principle any object, presented via any sense modality, at all). And please set aside the urge you are right now feeling to think of some trick whereby a purely symbolic T2 system could get around the problem of dealing with objects slipped in along with its symbols. (I will return to the subject of symbol systems later in this chapter.) The point is that trickery is not the issue here. We are discussing the T-hierarchy (T2-T4—I'll get to T1 shortly) as an empirical matter, and the candidate must be able to handle every robotic contingency that we can ourselves handle.

REVERSE ENGINEERING

I hope you are by now getting the sense that T3 is a pretty tall order for a machine to fill (not that T2 was not quite a handful already). Far from being a call for trickery, successfully generating T3 capacity is a problem in "reverse engineering," as Dan Dennett[3] has aptly characterized cognitive science. Determining what it takes to pass T3 is an empirical problem, but it is not an empirical problem in one of the basic natural sciences like physics or chemistry, the ones that are trying to discover the fundamental laws of nature. It is more like the problems in the engineering sciences, which *apply* those laws of nature so as to build systems with certain structural and functional capacities, such as bridges, furnaces, and airplanes; only in cognitive science the engineering must be done in reverse: The systems have already been built (by the Blind Watchmaker, presumably), and we must figure out what causal mechanism gives them their functional capacities. One way to do this is to build or simulate systems with the same functional capacities.

Well, if it is clear how the direct engineering problem of building a system that can fly is not a matter of trickery—not a matter of building something that *fools* us into thinking it is flying, whereas in reality it is not flying—then it should be equally clear how the reverse engineering problem of generating our T3 capacities is not a matter of trickery either: We want a system that can *really* do everything we can do, Turing Indistinguishably from the way we do it, not just something that fools us into thinking it can. In other words, the T tests are answerable both to our intuitions and to the *empirical constraints on engineering possibilities.*

Could a successful T3 candidate—a robot that walketh Turing-Indistinguishably among us for a lifetime, perhaps one or more of the "people" in this room right now—could such a successful candidate be a trick? Well, a trick in what sense?" one must at this point ask. That it does have full T3 capacity (and that this capacity is autonomous, rather than being, say, telemetered by another person who is doing the real work—the latter really *would* be cheating) we are here assuming as given (or rather, as an empirical engineering problem that we have somehow already solved successfully). Hence that cannot be the the the basis for any suspicions of trickery—any more than a plane with flight capacities indistinguishable from those of a DC-11 (assuming, again, that it is autonomous, rather than guided by an elaborate external system of, say, magnets) can be suspected of flying by trickery. Intuitions may differ as to what to call a plane with an *internal* system of magnets that could do everything a DC-11 could do by magnetic attraction and repulsion off distant objects. Perhaps that should be described as another form of flight, perhaps as something else, but what should be clear here is that the *task* puts some very strong constraints on the class of potentially successful candidates, and these constraints are engineering constraints, which is to say that they are *empirical* constraints.

UNDERDETERMINATION

This is the point to remind ourselves of the general problem of underdetermination of theories and models, both in the basic sciences and in engineering, direct and reverse: There is never any guarantee that any empirical theory that fits all the data—i.e., predicts and causally explains them completely—is the *right* theory. Here, too, we do not speak about trickery, but of degrees of freedom: Data constrain theories, they cut down on the viable options, but they do not necessarily reduce them to zero. More than one theory may successfully account for the same data; more than one engineering system may generate the same capacities. Since we are assuming here that the empirical scope of all the alternative rival candidates is the same—that they account for all and only the same data, that they have all and only the same capacities—the sole remaining constraints are those of economy: Some of the candidates may be simpler, cheaper, or what have you.

But economy is not a matter of trickery or otherwise either. Perhaps we should not have spoken of trickery, then, but of "reality." In physics, we want to know which theory is the "real" theory, rather than some lookalike that can do the same thing, but does not do it the way that Mother Nature happens to have done it.[26] Let us call this the problem of normal underdetermination: If we speak of complete or "Utopian" theories, the ones that successfully predict and explain *all* data—past, present, and future—there really is no principled way to pick among the rival candidates: Economy might be one, but Mother Nature might have been a spendthrift. You just have to learn to live with normal underdetermination in physics. In practice, of course, since we are nowhere near Utopia, it is the data themselves that cut down on the alternatives, paring down the candidates to the winning subset that can go the full empirical distance. (I will return to this.)

What are the data in that branch of reverse engineering called cognitive science? We have already lined them up: They correspond to the T-hierarchy I have been referring to. The first level of this hierarchy, T1, consists of "toy" models, systems that generate only a subtotal fragment or subset of our total capacity.[1] Being subtotal, T1 has a far greater degree of underdetermination than necessary. In pre-Utopian physics, every theory is T1 until all the data are accounted for, and then the theory has scaled up directly to T5, the Grand Unified Theory of Everything. The T2–T4 range is reserved for engineering. In this range, there are autonomous systems, subsets of the universe, such as organisms and machines, with certain circumscribed functional capacities. In conventional forward engineering, we stipulate these capacities in advance: We want a bridge that can span a river, a furnace that can heat a house, or a rocket that can fly to the moon. In reverse engineering these capacities were selected by the Blind Watchmaker: organisms that can survive and reproduce, fish that can swim, people that can talk. Swimming and

---

[1] Or, according to some more hopeful views, perhaps some autonomous, self-contained modular subcomponents of it.

talking, however, being subtotal and hence T1,[2] have unacceptably high degree of freedom. They open the door to ad hoc solutions that lead nowhere. In forward engineering, it would be as if flight engineers decided to model only a plane's capacity to fall, hop, or coast on the ground: there *might* be hints there as to how to successfully generate flight, but more likely T1 would send you off on countless wild-goose chases, none of them headed for the goal. In forward engineering the goal is T3: human "robotic" capacity; its direct-engineering homologue would be the flight capacity of a DC-11.

There is room for further calibration in forward engineering, too. Consider the difference between the real DC-11 and its T3-Indistinguishable (internal) magnetic equivalent. They are T3 equivalent, but T4-distinguishable, and for economic reasons we may prefer the one or the other. T2, in this case, would be a computer simulation of an airplane and its aeronautic environment, a set of symbols that was systematically interpretable as a plane in flight, just as a lifelong exchange of symbols with a T2 pen-pal is systematically interpretable as a meaningful correspondence with a person with a mind. We clearly want a plane that can really fly, however, and not just a symbol system that can be interpreted as flying, so the only role of T2 in aeronautic engineering is in "virtual world" testing of symbolic planes prior to building real T3 planes (planes with the full "robotic" capacities of real planes in the real air).[3]

Now what about reverse engineering and its underdetermination problems? Here is a quick solution (and it could even be the right one, although I am betting it is not, and will give you my reasons shortly). The reverse engineering case could go *exactly* as it does in the case of forward engineering: You start with T1, modeling toy fragments of an organism's performance capacity, then you try to simulate its total capacity symbolically (T2) and to implement the total capacity robotically (T3), while making sure the candidate is also indistinguishable neuromolecularly (T4) from the organism you are modeling, in other words, totally indistinguishable from it in all of its internal structures and functions.

Why am I betting this is not the way to go about the reverse engineering of the mind? Well, first, I am impressed by the empirical difficulty of achieving T3 at all, even without the extra handicap of T4 constraints, especially in the human case. Second, empirically speaking, robotic modeling (T3) and brain modeling (T4) are currently, and for the foreseeable future, independent data-domains. Insofar as T3 is concerned, the data are already in: We already know pretty much what it is that people can do; they can discriminate, manipulate, categorize, and discourse about the objects, events, and states of affairs in the world roughly the way we can; our task is now to model that capacity. Now T3 is a subset of T4; in other

[2] Unless we luck out and they turn out to be autonomous modules that can be veridically modeled in total isolation from all other capacities—a risky methodological assumption to make *a priori* in my opinion.

[3] There is, of course, also room for planes containing or telemetrically connected to T2 modules for computer-assisted "smart" flight, but that is another matter

words, what people can do is a subset of what their brains can do, but there is a wealth of structural and functional detail about the brain that may not only go well beyond its robotic capacity, but well beyond what we so far know about the brain empirically. In other words, there is no *a priori* way of knowing how much of all that T4 detail is *relevant* to the brain's T3 capacity; worse yet, nothing we have learned about the brain so far, empirically or theoretically, has helped us model its T3 capacities.[4]

So since (1) the T3 data are already in whereas the T4 data are not, since (2) the T4 data may not all be relevant to T3, and may even overconstrain it, and since (3) neither T4 data nor T4 theory has so far helped us make any headway on T3, I think it is a much better strategy to assume that T3 is the right level of underdetermination for cognitive modeling. This conclusion is strengthened by the fact that (4) T3 also happens to be the right version of the Turing Test, the one we use with one another in our practical, everyday "solutions" to the other-minds problem. And one final piece of support may come from the fact that (5) the Blind Watchmaker is no mind-reader either, and as blind to functionally equivalent, Turing-Indistinguishable differences as we are, and hence unable to favor one T3-equivalent candidate over another.[5]

THE SYMBOL GROUNDING PROBLEM

There are methodological matters here on which reasonable people could disagree; I am not claiming to have made the case for T3, the robotic level, over T4, the neural level, unassailably. But what about T2? There are those who think even T3 is too low a level for mind-modeling.[4,34,35] They point out that most if not all of cognition has a systematic, language-of-thought-like property that is best captured by pure symbol systems.[5,6] If you want a model to keep in mind as I speak of symbol systems, think of a natural language like English, with its words and its combinatory rules, or think of formal arithmetic, or of a computer programming language. A symbol system is a set of objects, called "symbol tokens" (henceforth just "symbols") that can be rulefully combined and manipulated, and the symbol combinations can be systematically interpreted as meaning something, as being *about* something. If your model for a symbol system was English, think of words and sentences, such as "the cat is on the mat," and what they can be interpreted as

[4] This is one of those controversial points I promised you, once we began fleshing out the T-hierarchy. I will speak about so-called "brain-style computation" and "neural nets" separately below; it is not at all clear yet whether these are fish or fowl, i.e., T2 or T4, whereas it is actually T3 we are after.

[5] Especially if the capacities to survive and reproduce are counted among our robotic capacities, as they surely ought to be—although this may bring in some molecular factors that are more T4 than T3.[23,27]

meaning; for arithmetic, think of expressions such as "1 + 1 = 2" and for computing think of the English propositions or arithmetic expressions formulated in your favorite programming language.

The "shape" of the objects in a symbol system (think of words and numerals) is arbitrary in relation to the "shape" of the objects, events, properties, or states of affairs that they can be systematically interpreted as being about: The symbol "cat" neither resembles nor is causally connected to the object it refers to; the same is true of the numeral "3." The rules for manipulating and combining the symbols in a symbol system are called "syntactic" because they operate only on the arbitrary shapes of the symbols, not on what they mean: The symbols' meaning is something derived from outside the symbol system, yet a symbol system has the remarkable property that it will bear the weight of a systematic interpretation. Not just any old set of objects, combined any old way, will bear this weight.[24] Symbol systems are a small subset of the combinatory things that you can do with objects, and they have remarkable powers: It was again Alan Turing, among others, who worked out formally what those powers were.[31,38] They amount to the power of computation, the power to express and do everything that a mathematician would count as doing something—and, in particular, anything a machine could do. The (Universal) Turing Machine is the archetype not only for the computer, but for any other machine, because it can symbolically emulate any other machine. It was only natural, then, for Turing to suppose that we ourselves were Turing machines.

The initial success of Artificial Intelligence (AI) seemed to bear him out, because, unlike experimental psychologists, unlike even behaviorists,[2] whose specialty was predicting, controlling, and explaining human behavior, AI researchers actually managed to *generate* some initially impressive fragments of behavior with symbol systems: chess playing, question-answering, scene-describing, theorem-proving. Although these were obviously toy models (T1), there was every reason to believe that they would scale up to all of cognition, partly because they were, for a while, the only kind of model that worked, and partly because of the general power of computation. Then there was computation's systematic, language-of-thought-like property.[5] And there was also the intuitive and methodological support from Turing's arguments for T2.

Symbol systems also seemed to offer some closure on the problem I raised at the beginning of this chapter, the mind/body problem, for a symbol system's properties reside at the formal level: the syntactic rules for symbol manipulation. The symbols themselves are arbitrary objects, and can be physically realized in countless radically different ways—as scratches on paper rulefully manipulated by people, as holes on a machine's tape, as states of circuits in many different kinds of computer computer—yet these could all be implementations of the same symbol system. So if the symbolists' hypothesis is correct, that cognition is computation and, hence, that mental states are really just implemented symbolic states, then it is no wonder that we have a mind/body problem in puzzling over what it might be about a physical state that makes it a mental state: for there is nothing special about the physical state except that it implements the right symbol system. The physical details are irrelevant. A radically different physical system implementing the same symbol system would be implementing the same mental state. (This is known in computer science as the hardware independence of the software and virtual levels of description.[30])

It was only natural, given all this, to conclude that any and every implementation of the symbol system that could pass T2 would have a mind. And so it was believed (and so it is still believed by many), although a rather decisive refutation of this hypothesis exists: John Searle[36] has pointed out the simple fact (based on the properties of symbols, syntax, and implementation-independence) that we would be wrong to conclude that a symbol system that could pass T2 in Chinese would understand Chinese, because Searle himself could become another implementation of that same symbol system (by memorizing and executing its symbol-manipulation rules) without understanding a word of Chinese. Searle could be your lifelong pen-pal without ever understanding a word you said. Now this, unlike prejudices about what machines and brains are and are not, *would* be a nonarbitrary reason for revising your beliefs about whether your lifelong pen-pal had really had a mind, about whether anyone in there had really been understanding what you were saying.

Not that I think you would ever have to confront such an awkward dilemma, for there are good reasons to believe that a pure symbol system could never pass T2[13]: Remember the problem of the photo included with your letter to your pen-pal. A picture is worth more than a thousand words, more than a thousand symbols, in other words. Now consider all the potential words that could be said about all the potential objects, events, and states of affairs you might want to speak about. According to the symbolist hypothesis, all these further symbols and symbol combinations can be anticipated and generated by prior symbols and the syntactic rules for manipulating them.[28] I have likened this to an attempt to learn Chinese from a Chinese/Chinese dictionary. It seems obvious that if you do not know Chinese already, then all you can do is go round and round in meaningless symbolic circles this way. To be sure, your quest for a definition would be as systematically meaningful *to someone who already knew Chinese* as the letter from your pen-pal would be, but the locus of that meaningfulness would not be the symbol system in either case, it would be the mind of the interpreter. Hence, it would lead to an infinite regress if you supposed that the mind of the interpreter was just a symbol system, too.

I have dubbed this the "symbol grounding problem"[14]: The meanings of the symbols in a pure symbol system are *ungrounded*; they are systematically interpretable by a system with a mind, but the locus of that interpretation is the mind of the interpreter, rather than the symbol system itself (just as the meaning of a book is in the minds of its readers, rather than in the book: the book is merely a bunch of symbols that can be systematically interpreted as meaningful by systems with minds). So if we are any kind of symbol system at all, we are surely *grounded* symbol systems, because the symbols in our heads surely do not mean what they mean purely in virtue of the fact that they are interpretable as meaningful by still other heads!

GROUNDING SYMBOLS IN ROBOTIC CAPACITY

How to ground symbols? How to connect them to the objects, features, events, and states of affairs that they are systematically interpretable as being about, but without the mediation of other minds? An attempt to find a system that is immune to Searle's argument already suggests an answer, for Searle's argument only works against pure symbol systems and T2. The moment you move to T3, Searle's "periscope," the clever trick that allowed him to penetrate the normally impenetrable other-minds barrier and confirm that there was no one home in there, namely, the *implementation independence of computation*, fails, and all candidates are again safe from Searle's snooping: For robotic T3 properties, unlike symbolic T2 properties, are not implementation-independent, starting from the most elementary of them, sensory and motor transduction. Searle could manage to *be* everything the T2 passing computer was (namely, the implementation of a certain symbol system) while still failing to understand, but there is no way to *be* a T3 robot without, among other things, *being* its optical transducers and its motor effectors, while failing to see or move (unless you implement only the part that comes *after* the transducer, but then you are not being the whole system and all bets are off).

So I take immunity to Searle's argument to be another vote for T3; but that is only the beginning. Symbol grounding requires a direct, unmediated connection between symbol and object.[20] Transduction is a good first step, but clearly the connection has to be selective, since not all symbols are connected to all objects. This brings us to the problem—and I emphasize that it is a T3 problem—of categorization.[12] We need a robot that can pick out and assign a symbolic name to members of object categories based on invariant features in their transducer projections—the shadows that objects cast on our sense organs. The robot must be able to categorize everything Turing-indistinguishably from the way we do.

Now I said I would come back to the question of neural nets and "brain style" computation. There is no space here to give my critique of (shall we call it) "hegemonic" connectionism or general complex or chaotic systems theories—the kind that want to take over all of cognition from symbol systems and do it all on their own. If and when they make significant inroads on the T3, such models will have commanded our attention; until then they are tools, just like everything else, including symbol systems.[25] Nor are neural nets in any realistic sense "brain-like," for the simple reason that no one really knows what the brain is like (T4 is even further from our empirical reach than T3). Indeed, connectionist modelers often unwittingly play a double game, offering hybrid tinker toys (T1) whose performance limitations (T3) are masked by their spurious neurosimilitude and whose lack of true brain-likeness (T4) is masked by their toy performance capacities.[17,22] Probably better to keep T3 and T4 criteria separate for now.

On the other hand, as items in the general cognitive armoury, neural nets are naturals for the important toy task of learning the invariants in the sensory projections of objects that allow us to sort them and assign them symbolic category names. For once you have such elementary symbols, grounded in the robotic capacity to discriminate, identify, and manipulate the objects they refer to, you can go on to combine them into symbol strings that define still further symbols and describe further objects and states of affairs. This amounts to breaking out of the Chinese/Chinese dictionary-go-round by giving the "dictionary" itself the capacity to pick out the objects its symbols are interpretable as being about, without any external mediation.[15,16]

For example, if you had looked up "ban-ma" in the Chinese/Chinese dictionary, you would have found it defined (in Chinese) as "striped horse," but that would be no help unless "striped" and "horse" were already grounded, somehow. My hypothesis is that grounding consists in the constraint exerted on the otherwise arbitrarily shaped symbols "horse" and "striped" by the nonarbitrary shapes of the analog sensory projections of horses and stripes and by the neural nets that have learned to filter those projections for the invariant features that allow things to be reliably called horses and stripes. The analog and feature-filtering machinery that connects the symbol to the projections of the objects it refers to exerts a functional constraint both on (1) the permissible combinations that those symbols can enter into, a constraint over and above (or, rather, under and below) the usual Boolean syntactic constraints of a pure symbol system, and on (2) the way the robot "sees" the world after having sorted, labelled, and described it in that particular way. The shape of the robot's world is *warped* by how it has learned to categorize and describe the objects in it.

CATEGORICAL PERCEPTION

I will now try to illustrate briefly the kind of dual analog/symbolic constraint structure that I think may operate in grounded hybrid symbol systems. The best illustration of the "receptive field" of an innate invariance filter is the red/green/blue feature detector in our color receptor system. Three kinds of units are selectively tuned to certain regions of the visual spectrum. Every color we see is then a weighted combination of the activity of the three, very much like linear combinations of basis vectors in a Euclidean vector space. This invariance filter is innate. One of its side effects (though this is not the whole story, and indeed the whole story is not yet known[39]), is what is called categorical perception (CP), in which the perceived differences within a color category are compressed and the perceived differences between different color categories are expanded.[12] This "warping" of similarity space makes members of the same color category look quantitatively and even qualitatively more alike, and members of different categories more different, than one would predict from their actual physical differences. If color space were not warped, it would to be a graded quantitive continuum, like shades of gray.

Color CP happens to be mostly innate, but CP effects can also be induced by learning to categorize objects in one way rather than another. With my collabarators at Vassar College, Janet Andrews and Kenneth Livingston, we were able to generate CP effects by teaching subjects, through trial and error with feedback, to sort a set of computer-generated Mondrian-like textures as having been painted by one painter or another (based on the presence of a subtle invariant feature that we did not describe explicitly to the subjects). The textures were rated for their similarity to one another by the subjects who had learned the categorization and by control subjects who had seen the textures equally often, but without knowing anything about who had painted what. Categorization compressed within category differences and expanded between category differences.[1]

In an analogous computer simulation experiment,[19,29] we presented to backpropagation neural nets twelve lines varying in length; the nets had to learn to categorize them as "short," "medium," and "long." Compared to control nets that merely performed auto-association (responding with a line that merely matches the length of the input), the categorization nets, like the human subjects, showed CP, with within-category compression and between-category expansion. The locus of the warping could be seen in how the "receptive fields" of the three hidden units changed as a result of categorization training, and the "evolution" of the warping throughout the training could be traced across time in hidden unit space. The warping occurs because of the way such nets succeed in learning the categorization: First, during autoassociation, the hidden-unit representations of each of the lines move as far away from one another as possible; with more analog inputs, this tendency is constrained by their analog structure; categorization is accomplished by partitioning the cubic hidden unit space into three regions with separating planes. So the compression/dilation occurs because of the way the hidden-unit representations must change their locations in order to get on the correct side of this plane, while still being constrained by their analog structure, and with the separating planes exerting a "repulsive" force that is strongest at the category boundaries.

This is the kind of nonarbitrary shape constraint that would be "hanging" from every ground-level symbol in a grounded hybrid symbol system and would be inherited by the higher-level symbols defined in terms of the ground-level symbols. A "thought" in the head of such a robot would then not just be the activitation of a string of symbols, but of all the analog and invariance-detecting machinery to which the symbols were connected, machinery which would, in turn, ground them in the objects they were about.

UNDERDETERMINATION REDUX

Now we come to the question raised in the title of this chapter: Does mind piggyback on our robotic and symbolic capacities? My reply is that all we can do is hope so, because there will never be any way to know better. Even in a grounded robot—one that is T3-indistinguishable from us, immune to Searle's argument, able to discriminate, identify, manipulate, and discourse about the objects, events, and states of affairs that its symbols are systematically interpretable as being about, and able to do so without the mediation of any outside interpreter, thanks to the bottom-up constraints of its grounding—it is still possible that there is no one home in there, experiencing experiences, no one that the duly grounded symbols are about what they are about *to*. Just a mindless "Zombie."[26]

If this Zombie possibility were actually realized, then we would have to turn to T4 to fine-tune the robot by tightening our Turing filter still further. But I have to point out that, without some homologue of Searle's periscope, there would be no way we could *know* that our T3-scale robot was just a Zombie, that T3 had hence been *unsuccessful* in generating a mind, and that we accordingly needed to scale up to T4. Nor would we have any way of knowing that there was indeed someone home in a T4 candidate either. I am not inclined to worry about that sort of thing, however; in fact, as far as I am concerned, T3 is close enough to Utopia so that you can call in the hermeneuticists at that point and mentally interpret it to your heart's content. I do not believe that a Zombie could slip through an empirical filter as tight as that. It is only the premature mentalistic interpretation of subtotal T1 toys or of purely symbolic T2 modules that I would caution against, for overinterpretation will invariably camouflage excess underdetermination, just as premature neurologizing does.

The degrees of underdetermination of mental modeling will always be greater than those of physical modeling, however, regardless of whether we prefer T3 or we hold out for T4, because believing we've captured the mental will always require a leap of faith that believing we've captured the physical does not. This has nothing at all to do with unobservability; quarks are every bit as unobservable as qualia.[26] But, unlike qualia, quarks are allowed to do some independent work in our explanation. Indeed, if quarks still figure in the Utopian Grand Unified Theory of Everything (T5), they will be formally indispensable; remove them and the theory no longer predicts and explains. But for the Utopian Theory of Mind, whether T3 or T4, the qualia will always be a take-it-or-leave-it hermeneutic option, and I think that's probably because allowing the qualia any independent causal role of their own would put the rest of physics at risk.[10,27] So even if God could resolve the remaining indeterminacy, assuring us that our theory was not only successful and complete, but also the *right* theory, among all the possibilities that underdetermination left open and hence that our hermeneutics was correct too, this, like any other divine revelation, would *still* call for a leap of faith on our part in order to believe. This extra element of underdetermination (and whatever perplexity and dissatisfaction it engenders) will remain the unresolvable residue of the mind/body problem.

REFERENCES

1. Andrews, J., K. Livingston, S. Harnad, and V. Fischer. "Learned Categorical Perception in Human Subjects: Implications for Symbol Grounding." In preparation.
2. Catania, A. C., and S. Harnad, eds. *The Selection of Behavior. The Operant Behaviorism of B. F. Skinner: Comments and Consequences.* New York: Cambridge University Press, 1988.
3. Dennett, D. C. "Cognitive Science as Reverse Engineering: Several Meanings of 'Top Down' and 'Bottom Up.'" In *Proceedings of the 9th International Congress of Logic, Methodology and Philosophy of Science,* edited by D. Prawitz, B. Skyrms, and D. Westerstahl. North Holland: in press.
4. Dietrich, E. "Computationalism." *Soc. Epis.* **4** (1990): 135–154.
5. Fodor, J. A. *The Language of Thought.* New York: Thomas Y. Crowell, 1975.
6. Fodor, J. A., and Z. W. Pylyshyn. "Connectionism and Cognitive Architecture: A Critical Appraisal." *Cognition* **28** (1988): 3–71.
7. Harnad, S., H. D. Steklis, and J. B. Lancaster, eds. *Origins and Evolution of Language and Speech,* 280.. New York: NYAS, 1976.
8. Harnad, S., R. W. Doty, L. Goldstein, J. Jaynes, and G. Krauthamer, eds. *Lateralization in the Nervous System.* New York: Academic Press, 1977.
9. Harnad, S. "Neoconstructivism: A Unifying Theme for the Cognitive Sciences." In *Language, Mind and Brain,* edited by T. Simon and R. Scholes, 1–11. Hillsdale, NJ: Erlbaum, 1982.
10. Harnad, S. "Consciousness: An Afterthought." *Cogn. & Brain Theory* **5** (1982): 29–47.
11. Harnad, S. "What Are the Scope and Limits of Radical Behaviorist Theory?" *Behav. & Brain Sci.* **7** (1984): 720–721.
12. Harnad, S., ed. *Categorical Perception: The Groundwork of Cognition.* New York: Cambridge University Press, 1987.
13. Harnad, S. "Minds, Machines and Searle." *J. Theor. Exper. Art. Int.* **1** (1989): 5–25.
14. Harnad, S. "The Symbol Grounding Problem." *Physica D* **42** (1990): 335–346.
15. Harnad, S. "Against Computational Hermeneutics. (Invited Commentary on Eric Dietrich's Computationalism)." *Soc. Epis.* **4** (1990): 167–172.
16. Harnad, S. "Lost in the Hermeneutic Hall of Mirrors. Invited Commentary on: Michael Dyer: Minds, Machines, Searle and Harnad." *J. Exper. Theor. Art. Int.* **2** (1990): 321–327.
17. Harnad, S. "Symbols and Nets: Cooperation vs. Competition." *Conn. & Sym. Conn. Sci.* **2** (1990): 257–260.
18. Harnad, S. "Other Bodies, Other Minds: A Machine Incarnation of an Old Philosophical Problem." *Minds & Machines* **1** (1991): 43–54.

19. Harnad, S., S. J. Hanson, and J. Lubin. "Categorical Perception and the Evolution of Supervised Learning in Neural Nets." In *Working Papers of the AAAI Spring Symposium on Machine Learning of Natural Language and Ontology*, edited by D. W. Powers and L. Reeker, 65–74. Presented at Symposium on Symbol Grounding: Problems and Practice, Stanford University, March 1991; also reprinted as Document D91-09, Deutsches Forschungszentrum fur Kuenstliche Intelligenz GmbH Kaiserslautern FRG, 1991.

20. Harnad, S. "Connecting Object to Symbol in Modeling Cognition." In *Connectionism in Context*, edited by A. Clarke and R. Lutz. Berlin: Springer-Verlag, 1992.

21. Harnad, S. "The Turing Test is Not a Trick: Turing Indistinguishability is a Scientific Criterion." *SIGART Bull.* **3(4)** (1992): 9–10.

22. Harnad, S. "Grounding Symbols in the Analog World with Neural Nets." Special Issue on "Connectionism versus Symbolism," edited by D.M.W. Powers & P.A. Flach. *Think* **2** (1993): 12–78.

23. Harnad, S. "Artificial Life: Synthetic Versus Virtual." In *Artificial Life III*, edited by C. G. Langton. Santa Fe Institute Studies in the Sciences of Complexity, Proc. Vol. XVI. Reading, MA: Addison-Wesley, 1993.

24. Harnad, S. "The Origin of Words: A Psychophysical Hypothesis." In *Muenster: Nodus Pub*, edited by W. Durham and B. Velichkovsky. [Presented at Zif Conference on Biological and Cultural Aspects of Language Development. January 20–81522, 1992 University of Bielefeld], 1993.

25. Harnad, S. "Symbol Grounding is an Empirical Problem: Neural Nets are Just a Candidate Component." In *Proceedings of the Fifteenth Annual Meeting of the Cognitive Science Society*. Earlbaum, NJ: Erlbaum, 1993.

26. Harnad S. "Discussion (passim)." In *Experimental and Theoretical Studies of Consciousness*, edited by G. R. Bock and J. Marsh. CIBA Foundation Symposium 174. Chichester: Wiley, 1993.

27. Harnad, S. "Turing Indistinguishability and the Blind Watchmaker." Paper presented at London School of Economics Conference of "Evolution and the Human Sciences," June 1993.

28. Harnad, S. "Problems, Problems: The Frame Problem as a Symptom of The Symbol Grounding Problem." *PSYCOLOQUY* **4(34)** (1993): 11.

29. Harnad, S., S. J. Hanson, and J. Lubin. "Learned Categorical Perception in Neural Nets: Implications for Symbol Grounding." In *Symbol Processing and Connectionist Network Models in Artificial Intelligence and Cognitive Modelling: Steps Toward Principled Integration*, edited by V. Honavar and L. Uhr, 1994.

30. Hayes, P., S. Harnad, D. Perlis, and N. Block. "Virtual Symposium on Virtual Mind." *Minds & Machines* **2** (1992): 217–238.

31. Lewis, H., and C. Papadimitriou. *Elements of the Theory of Computation*. Englewood Cliffs, NJ: Prentice Hall, 1981.

32. Nagel, T. "What is It Like to Be a Bat?" *Phil. Rev.* **83** (1974): 435–451.

33. Nagel, T. *The View From Nowhere*. New York: Oxford University Press, 1986.
34. Newell, A. "Physical Symbol Systems." *Cog. Sci.* **4** (1980): 135–183.
35. Pylyshyn, Z. W. *Computation and Cognition*. Cambridge MA: MIT/Bradford, 1984.
36. Searle, J. R. "Minds, Brains, and Programs." *Behav. & Brain Sci.* **3** (1980): 417–424.
37. Turing, A. M. "Computing Machinery and Intelligence." In *Minds and Machines*, edited by A. Anderson. Engelwood Cliffs, NJ: Prentice Hall, 1964.
38. Turing, A. M. *Mechanical Intelligence*, edited by D. C. Ince. North Holland, 1990.
39. Zeki, S. *Colour Vision and Functional Specialisation in the Visual Cortex*. Amsterdam: Elsevier, 1990.

Daniel C. Dennett
Tufts University, Center for Cognitive Studies, Medford, MA 02155

Evolution as An Algorithm—The Ultimate Insult?

John Locke[3] offered what he considered a sound *a priori* argument that Mind must come first, must be the original Cause, not merely an Effect:

> If, then, there must be something eternal, let us see what sort of Being it must be. And to that it is very obvious to Reason, that it must necessarily be a cogitative Being. For it is as impossible to conceive that ever bare incogitative Matter should produce a thinking intelligent Being, as that nothing should of itself produce Matter. Let us suppose any parcel of Matter eternal, great or small, we shall find it, in itself, able to produce nothing...Matter then, by its own strength, cannot produce in itself so much as Motion: the Motion it has, must also be from Eternity, or else be produced, and added to Matter by some other Being more powerful than Matter...But let us suppose Motion eternal too: yet Matter, incogitative Matter and Motion, whatever changes it might produce of Figure and Bulk, could never produce Thought: Knowledge will still be as far beyond the power of Motion and Matter to produce, as Matter is beyond the power of nothing or nonentity to produce. And I appeal to everyone's own thoughts, whether he cannot as easily conceive Matter produced by nothing, as Thought produced by pure Matter, when before there was no

such thing as Thought, or an intelligent Being existing...So if we will suppose nothing first, or eternal: Matter can never begin to be: If we suppose bare Matter, without Motion, external: Motion can never begin to be: If we suppose only Matter and Motion first, or eternal: Thought can never begin to be. For it is impossible to conceive that Matter either with or without Motion could have originally in and from itself Sense, Perception, and Knowledge, as is evident from hence, that then Sense, Perception, and Knowledge must be a property eternally inseparable from Matter and every particle of it.

Darwin's great contribution was the complete destruction of Locke's argument and the nearly universal blockade of imagination that it both reflected and fostered. An early critic of Darwinian thinking put his finger on it:

In the theory with which we have to deal, Absolute Ignorance is the artificer; so that we may enunciate as the fundamental principle of the whole system, that, IN ORDER TO MAKE A PERFECT AND BEAUTIFUL MACHINE, IT IS NOT REQUISITE TO KNOW HOW TO MAKE IT. This proposition will be found, on careful examination, to express, in condensed form, the essential purport of the Theory, and to express in a few words all Mr. Darwin's meaning; who, by a strange inversion of reasoning, seems to think Absolute Ignorance fully qualified to take the place of Absolute Wisdom in all the achievements of creative skill.[1]

Exactly! What this critic saw and hated is precisely what others have seen and loved: the idea (whether or not Darwin saw it with clarity) that all the Design in the universe can be explained as the product of a process that is ultimately bereft of intelligence, in other words, an algorithmic process that weds randomness and selection to produce a unique, branching, recursively revised structure that bears as its fruit all the intelligence that exists. This "strange inversion of reasoning," as the critic calls it, spreads its revolutionary perspective through all of science, unifying as it reformulates most of the philosophical ideas that have guided (and often misguided) science.

The anonymous critic has had many intellectual descendants, up to the present. A recent critic puts the same misgivings as follows:

Life on Earth, initially thought to constitute a sort of prima facie case for a creator, was, as a result of Darwin's idea, envisioned merely as being the outcome of a process and a process that was, according to Dobzhansky, "blind, mechanical, automatic, impersonal," and, according to de Beer, was "wasteful, blind, and blundering." But as soon as these criticisms [sic] were leveled at natural selection, the "blind process" itself was compared to a poet, a composer, a sculptor, Shakespeare—to the very notion of creativity that the idea of natural selection had originally replaced. It is clear, I think, that there was something very, very wrong with such an idea."[2]

Or something very, very right. Many who have shared Bethell's discomfort have sought to contain Darwinian thinking within some "proper" sphere—ceding it some limited role within biology, but trying to prohibit its spread into cosmology, psychology, the arts, ethics, and religion. These attempts at containment have largely been misguided, but as byproducts they have often produced important strengthenings and deepenings of the underlying Darwinian idea. Let us understand a *skyhook* to be a "mind first" force or power or process, an exception to the principle that all design, and apparent design, is ultimately the result of mindless, purposeless, mechanical processes. Then a *crane* is a subprocess or special feature of a design process that can be demonstrated to permit the local speeding up of the basic, slow process of natural selection, and that can be demonstrated to be itself the predictable (or retrospectively explicable) product of the basic process. Much of the most fruitful (if often extremely emotional and even vicious) controversy in evolutionary thinking since Darwin can then be characterized as searchers for skyhooks discovering cranes. Time and again, challenges to Darwinian thinking of the "you can't get here from there in the time available" variety have been met by discoveries or reformulations that show how the underlying Darwinian algorithmic processes can be cascaded recursively into ever more powerful and swift mechanisms for "lifting" in Design Space. The idea of evolution as fundamentally an algorithmic process is often misunderstood, but it is, in fact, the source of the power of Darwin's contribution. (The claims put forth in this abstract are discussed at length in my forthcoming book.)[4]

REFERENCES

1. Anonymous. Review in *Atheneum* **102** February 8, 1867: 217.
2. Bethell, T. "Darwin's Mistake." *Harper's Mag.* (1976): 70–75.
3. Locke, J. *Essay Concerning Human Understanding*, IV, x, 10. London, 1690.
4. Dennett, D. *Darwin's Dangerous Idea*. New York: Simon & Schuster, 1995.

Milton Keynes UK
Ingram Content Group UK Ltd.
UKHW020028071024
449327UK00032B/2968